More Praise for *The Future of Geography*

"An essential take on power, politics, and the future of humanity."
—*Daily Telegraph*

"Insightful . . . Looking to the stars shows that humanity's future will be far more grounded than speculative fiction film and television would suggest."
—*Diplomatic Courier*

"Thought-provoking and worrying in equal measures."
—*Popular Science*

"Marshall is an engaging writer, good at explaining the science as well as the politics, and with an eye for a telling fact."
—*The New Statesman*

"Well-researched . . . informative . . . Marshall displays flashes of wit and a thorough understanding of the issues."
—*Kirkus Reviews*

"A fascinatingly insightful book, written with humor and excitement about the road ahead. But it's also a cautionary tale of what could happen unless we can work towards greater international collaboration and transparency."
—*Geographical*

"[Marshall's] prose is brisk in pace and refreshingly crystalline in its clarity, affording a highly readable lesson in historical geopolitics."
—*BBC Sky at Night Magazine*

"Compelling . . . important . . . Marshall demonstrates a refreshing understanding of the new space age—one in which humanity will go into space and stay there. . . . It provides a practical vision of our necessary future in space."

—*Times Literary Supplement*

"Lead[s] the reader on a voyage of galactic discovery."

—*The Mail on Sunday*

"[An] engaging exploration of power politics in space."

—*Irish Independent*

"Thought-provoking . . . Marshall's assessments intrigue, and he provides valuable insight into an overlooked aspect of space exploration."

—*Publishers Weekly*

"A prophetic vision of what the geopolitics of space could look like over the next fifty years."

—ClimatewithBrian.com

"A gripping and amazing journey . . . guide[s] us expertly and often amusingly into the exploration of space and its profound implications for those of us left behind on planet Earth."

—Ian Goldin, professor of globalization and development at the University of Oxford

"A chilling, insightful exploration of the political and military implications of our presence in space."

—Brian Clegg, author of *Final Frontier*

"This stirring book shows why astropolitics matters for everyone on earth—not just the scientists and the dreamers."

—Dr. Bleddyn E. Bowen, author of *War in Space*

THE FUTURE

of

GEOGRAPHY

How the Competition in Space
Will Change Our World

Tim Marshall

SCRIBNER
New York   Amsterdam/Antwerp   London   Toronto   Sydney   New Delhi

Scribner
An Imprint of Simon & Schuster, LLC
1230 Avenue of the Americas
New York, NY 10020

First Scribner trade paperback edition February 2025

SCRIBNER and design are trademarks of Simon & Schuster, LLC

For information about special discounts for bulk purchases, please contact Simon &
Schuster Special Sales at 1-866-506-1949 or business@simonandschuster.com.

The Simon & Schuster Speakers Bureau can bring authors to your live event. For
more information or to book an event, contact the Simon & Schuster Speakers
Bureau at 1-866-248-3049 or visit our website at www.simonspeakers.com.

Illustration credits: pp. 8–9: NASA/JPL; pp. 28–29 and pp. 172–73: NASA;
pp. 56–57: NASA/Alex Gerst; pp. 62 and 68: courtesy of Elliott & Thompson;
pp. 78–79: NASA/Johns Hopkins APL/Steve Gribben; pp. 106–7: Shujianyang,
CC BY-SA 4.0; pp. 128–29: NASA/Terry White/SLS; pp. 150–51: NASA/Bill
Ingalls; pp. 204–5: Shutterstock; pp. 226–27: NASA/JPL-Caltech

Manufactured in the United States of America

1 3 5 7 9 10 8 6 4 2

Library of Congress Control Number: 2023025375

ISBN 978-1-6680-3164-3
ISBN 978-1-6680-3165-0 (pbk)
ISBN 978-1-6680-3166-7 (ebook)

To my family

CONTENTS

INTRODUCTION

I haven't been everywhere, but it's on my list.
—Susan Sontag

We explored the world and discovered it is finite. Now, just as our territory and resources begin to run out, we find that the big, beautiful ball in the sky—the Moon—is full of the minerals and elements we all need. It's also a launchpad: just as early humans went from island to island as they crossed the seas, so the Moon will allow us to reach across the solar system and beyond.

It's no surprise, then, that we are in a new Space Race. To the victor the spoils. The challenge will be to ensure that humanity is the victor.

Space has shaped human life from our very beginning. The heavens explained our early creation stories, influenced our cultures, and inspired scientific advances. But our view of space is changing. It is now, more than ever, becoming an extension of the geography of Earth. Humans are taking our nation-states, our corporations, our history, and our politics and conflicts way up above us. And that could revolutionize life down on Earth's surface.

Space has already changed much in our everyday lives. It is central to communication, economics, and military strategy, and increasingly important to international relations. It is now also becoming the latest arena for intense human competition.

The signs that space is going to be a huge geopolitical narrative of the twenty-first century have been accumulating for some time. In recent years, rare metals and water have been found on the Moon; private companies such as Elon Musk's SpaceX have massively lowered the cost of breaking through the atmosphere; and the big powers have fired missiles from Earth, blowing up their own satellites to test new weapons. All these events have been pieces of the bigger story emerging.

To understand that story, it is helpful to see space as a place with geography: it has corridors suited to travel, regions with key natural assets, land on which to build, and dangerous hazards to avoid. For the last few decades all of this was considered the common property of humanity—no sovereign nation could exploit or lay claim to any of it in its own name. But that idea, enshrined in several noble, albeit outdated and unenforceable documents, is fraying badly. The nations of Earth are all looking to take advantage where they can. Throughout recorded history, civilizations fortunate enough to be able to utilize natural resources have developed technologies to help themselves grow stronger, and eventually dominate others.

It doesn't have to be that way. We have many examples of cooperation in space, and many of the space-related technologies being developed, in medicine and clean energy for example, will help us all. Several countries are working on ways to deflect huge asteroids, capable of destroying the world, off a collision course—and it doesn't get more common property than that. As the science fiction writer Larry Niven said, "The dinosaurs became extinct because they didn't have a space program." It would be beyond inconvenient to suffer another hit like that.

It's taken a long time to get where we are. The Big Bang theory suggests that 13.7 billion years ago, give or take the odd few thousand years, every single thing in the universe that exists today was compressed into an infinitesimally tiny particle existing in nothingness. Some concepts related to the universe can be difficult to get your head around, and "nothingness" is one that scientists argue over endlessly. They go into notions such as quantum vacuums, in which ripples in space can cause things to pop into existence and back out again, but after reading and rereading the theories several times over, I'm never much further along. The universe is expanding—but into what? What is outside its current boundaries? I can't imagine nothing. An endless wall of gray does the trick (beige is also available), but only for a second because, of course, gray is something and not nothing . . . and then I give up. Fortunately, theoretical physicists and cosmologists are made of sterner stuff.

From "nothingness" the particle exploded—although it wasn't so much "flash, bang, wallop!" as "bang, wallop, flash!" as it took about 380,000 years for the first particles of light to emerge. This is the cosmic microwave background (CMB), which scientists can see through modern space telescopes—all the way back, almost to the very beginning. You can see it for yourself in the static fuzz between channels when you tune an old analog TV. The universe expanded and cooled, and gravity caused gas clouds to gather and condense into stars.

We now know that our Sun was formed roughly 4.6 billion years ago—a relative newcomer in the universe. A huge disc of gas and heavier debris swirling around the new star then created the planets and their moons in our solar system.

Planet Earth is the third rock from the Sun. It's a good place to be. In fact, for now it's the only place because if it was anywhere else, we wouldn't be. Everything that has happened since the Big Bang has shaped the geography of what we see now and allowed us to evolve to where we are. Earth is the Goldilocks of planets. Not too hot, not too cold—just right for life. The Earth's position, size, and atmosphere all contribute to keeping us grounded. Literally. Its size means gravity has enough strength to hold on to the atmosphere. Move elsewhere in our neck of infinity and we'd either fry, freeze, or suffocate due to a lack of breathable air.

As the great American cosmologist Carl Sagan said in his book *Billions & Billions*, "Many astronauts have reported seeing that delicate, thin, blue aura at the horizon of the daylit hemisphere—that represents the thickness of the entire atmosphere—and immediately, unbidden, began contemplating its fragility and vulnerability. They worry about it. They have reason to worry." You'd think we might take better care of it.

But humans have always been wanderers, and in the last century have begun to move far from our planet. Space is such a massive canvas that we have only sketched our presence on it in a tiny corner. The rest is there for us to draw on in detail—together. If we're to navigate our way outward into the next era of the Space Age in a peaceful and cooperative fashion, we need to understand space in its historical, political, and military contexts, and to grasp what it will mean for our future.

In these chapters, we will look back in time to see how space has influenced our cultures and our ideas, from societies organized largely around religion, all the way to scientific revolutions. From there, it was the Cold War that drove the Space Race—prompting huge leaps in human endeavor and innovation that finally allowed us to break the bonds of Earth. Once out, we started to see opportunities, resources, and strategic points worth competing for. We are now in the era of astropolitics. But what we've failed to establish so far is a set of universally agreed-upon rules to regulate this competition; without laws governing human activity in space, the stage is set for disagreements on an astronomical level.

In the modern era, there are three main players we need to know about: China, the US, and Russia. These are the independent spacefaring nations, and how they choose to proceed will affect everyone else on Earth. The militaries of each have a version of a "Space Force" that provides war-fighting capabilities for their forces on land, sea, and in the air. All are increasing their capacity to attack and defend the satellites that provide those capabilities.

The rest of the nations know they can't compete with the Big Three, but they still want to have a say in what goes up and what comes down; they are assessing their options and aligning into "space blocs." If we cannot find a way to move forward as a unified planet, there is an inevitable outcome: competition and possibly conflict played out in the new arena of space.

And finally, we'll look far forward into our future, to see what space could hold for us—on the Moon, on Mars, and beyond.

The Moon pulls the sea to the shore, and humans to its surface. Wolves raise their muzzles and howl at the silvery disc hanging in the night sky. Humans raise their eyes and look farther, to infinity. We always have, and now we are on our way.

PART I

The Path to the Stars

Our solar system

LOOKING UP

*To confine our attention to terrestrial matters
would be to limit the human spirit.*
—Stephen Hawking

The flickering lights of the stars tell many stories. Long before we ever dreamed of venturing into space, before artificial light dimmed our view, we stared up at the skies and asked, "Why is there something rather than nothing?" Much of human endeavor has been driven by our desire to reach the stars.

The first recorded beliefs about creation, the gods, and constellations most likely come from an oral storytelling tradition stretching back into prehistory. All ancient cultures saw in the sky an idea of what might have created them, who they were, what their role was, and how they should behave. If there were gods—and what else could explain what was seen?—it was logical to believe that some of them lived in the heavens above.

Humans are hardwired to look at things and see patterns. People joined the dots and made a picture corresponding to what they saw on Earth and what they knew from their legends. Those in hot climates might see the shapes of scorpions or lions, while those in colder realms would pick out a moose. In Finland the northern lights are known as "fox fires" because of the ancient tale of a magical fox whose tail swept snow up into the heavens, while in parts of Africa there is a legend that the Sun is behind the night sky and the stars are holes that let some of its light through. The stars were inseparable from our stories, myths, and legends.

The earliest potential evidence of people trying to analyze and understand the skies dates to about 30,000 years ago, toward the end of the last ice age. In the early 1960s, the prehistorian Alexander Marshack interpreted marks carved into animal bones as being lunar calendars. The bones show sequences of twenty-eight and twenty-nine points. Experts still argue about exactly what women and men in the Late Paleolithic period might have known, but there is a body of evidence that they were studying the stars.

Scientists speculate that these early astronomers used their portable calendars as they moved on long hunting trips and migrations, and possibly for rituals. It makes sense that a way of marking time would develop. You would need to know when, for example, the mosquito season was about to begin, or when you should move on toward the trees whose fruit was ripe.

The more practical side of watching the skies was also crucial as hunter-gatherers became more sedentary, a process that began roughly 12,000 years ago. The first farmers and herders needed to know when to sow seeds and how long it was before harvest. Some of the Neolithic cave paintings found in Europe, which are over 10,000 years old, are thought to depict star formations. Again, the claims are debated, but the patterns of constellations can be found in some of the animal drawings. People who looked at the stars every clear night must have noticed that the lights were in different positions at different times, even if they had not yet worked out that 365 periods of daylight and darkness equaled one unit of time.

We are still a long way from any proof of accurate measurement of the movement of the planets and stars at that time. Even when we arrive at the beginning of the building of stone circles, the evidence is sketchy.

The oldest known is Nabta Playa in what is now Egypt. It's sometimes called the Stonehenge in the Sahara, which is a bit unfair as it was built about 7,000 years ago, some 2,000 years before the world's most famous henge. This is because the site was only discovered in the 1970s and fully excavated in the 1990s. It's believed to have been built by seminomadic herders to help them know when they should be on the move. There's some evidence to suggest that the stones were aligned with key stars, such as Sirius, which is the brightest star in the night sky. Evidence for the more fanciful suggestion that they could also measure the distance to those stars is harder to find, mostly because, according to experts, it isn't there.

The same is true of Stonehenge and the many other stone circles in northwest Europe. Stonehenge was first constructed about 5,000 years ago, by which time farming had been a way of life in the region for 1,000 years. It is safe to say that Stonehenge lines up with the Sun on the winter and summer solstices, but beyond that any association with astronomy is far more speculative. It's known that great feasts were held near the monument from the thirty-eight thousand discarded animal bones found at a settlement almost two miles away. Alas, Druids are not thought to have been present at these events as they didn't show up in Britain until about 2,000 years later, which must be quite disappointing for those people who descend on the site today dressed in white gowns and carrying sticks.

It's when we reach back about 4,000 years that we begin to find written proof that people were analyzing the skies with a high level of sophistication and the ability to predict movements accurately. Writing and mathematics were the keys enabling the breakthrough.

In around 1800 BCE the Babylonians, borrowing from their predecessors, the Sumerians, wrote down the signs of the zodiac based on the constellations as they saw them. They had long believed that the gods sent them warnings from the sky about future events such as famine. Priests developed the ability to record celestial movements on clay tablets and designed a calendar featuring twelve lunar months. That was the relatively easy part. After a few generations of storing the data, and using advancements in mathematics, they noticed that planets do not move in the same way in consecutive years but, given long enough, patterns of repetition do occur. This allowed them to work out where in the sky a planet would be on a specific date in the future.

It's largely down to the Babylonians that we divide time into seven-day weeks. They saw seven celestial bodies, figured that each one oversaw a particular day, and so divided the lunar cycle of twenty-eight days into four parts. At the time, the Egyptians were using a ten-day division, which, had it lasted, would make for a long workweek. As for a two-day weekend? Well, the Babylonians did designate one day for relaxation, but we can also thank the Hebrews for letting us know that if God wanted to rest on the seventh day, then so should we. Somewhat later, the unions won us another day off whether God wanted one or not.

The Assyrians, Egyptians, and others made similar advances in astronomy, but humanity still believed that astronomical events were caused by the gods. Astronomy and astrology were inseparable. The ancient Greeks thought the same way as they took up the mantle of these scientific pioneers. The Greeks put their stamp on cosmology like no other civilization. By looking up at the stars, they also changed the way we think about the world.

The Greeks had been learning from the Babylonians for centuries. Pythagoras was just one of those who had benefited when, around 550 BCE, he worked out that what were called the morning star and the evening star were the same thing—the planet Venus. The breakthroughs he and others went on to achieve came as they applied geometry and trigonometry to cosmic questions.

One of the greats was Hipparchus, who is thought to have invented the astrolabe—Greek for "star-taker." This was the "smartphone" of the ancients and, unlike some of today's consumer technology, it didn't have a built-in failure date. Astrolabes were used for almost 2,000 years. They could tell you where you were, what time it was, when the Sun would set, and give you your horoscope. They functioned using a series of sliding plates, including ones containing Earth's latitudinal lines and the location of certain stars. They spread from the Hellenic Greek region into the Arab countries and later to Western Europe. The Muslims used them to locate the direction of Mecca; Columbus used one as he headed toward the Americas.

The Greeks believed Earth to be round several generations before Aristotle describes it as such in his *On the Heavens*, written in 350 BCE. He noted that Earth's shadow on the Moon during an eclipse is circular. If Earth was a flat disc, then at some point, when sunlight struck it side on, its shadow on the Moon would be a line. As this did not happen, logic suggested a round Earth.

Aristotle writes about mathematicians measuring distance in *stades* (from which we get the word "stadium") and finding that Earth's circumference was 400,000 *stadia*—about 45,500 miles. They may have been off by 20,000 miles, but it was still a massive leap forward in our thinking.

About a hundred years later, Eratosthenes of Cyrene worked out how to measure the circumference of Earth accurately. He knew of a well in Syene (now called Aswan), Egypt, where every year at the summer solstice the Sun illuminated the bottom of the well without casting any shadows. This meant the Sun was directly overhead. He then measured the length of the shadow cast by a stick at noon on the summer solstice in Alexandria. From this, he calculated that the difference in the Sun's elevation between the two cities equated to an angle of 7.2 degrees along the curvature of the Earth—roughly one-fiftieth of a circle. Now all he needed was an accurate measurement of the distance from Alexandria to Syene. He hired professional surveyors, trained to walk with equal strides, and was told the distance was 5,000 *stadia*. His conclusion was that Earth's circumference was between 25,010 and 28,521 miles. The actual circumference is now usually accepted as 24,914 miles.

At its heart, Greek learning argued that there is an underlying order to the universe and that this could be discovered and expressed by observation and mathematics. This was the beginning of the idea that the world could be understood through natural processes, rather than with reference to the gods. The Greeks worked to find the circumference of the Moon, and the distance from Earth to the Moon, and the Moon to the Sun. However, they consistently vastly underestimated distance and, although they developed theoretical models of planetary motion, in all of them the planets circled Earth, a belief that survived until the Renaissance.

There were many scientific giants, culminating with Claudius Ptolemy (*c.* 100–*c.* 170 CE), who summarized classical astronomy and categorized the star pictures of the ancients into forty-eight constellations (today there are eighty-eight), giving them names that still dominate many languages. Aquarius, Pegasus, Taurus, Hercules, Capricorn, etc., were all written down in Ptolemy's book, which he called *The Mathematical Compilation* but is known to the world by its Arabic name, the *Almagest.* Yet Ptolemy was hamstrung by the same thought process as his predecessors: that Earth was the center of the universe, and the planets circled it.

It was based on what they knew and what their logic told them, and this model held for more than 1,500 years. We know of one early exception to this orthodox view. Aristarchus of Samos (310–230 BCE) argued that Earth revolved around the Sun—the heliocentric universe model. The scholars disagreed.

Aristarchus and others had correctly worked out the distance to the Moon. However, they put the Sun only about twenty times farther away than that—a massive underestimation, but still an enormous distance. The Greeks erred on the side of caution. To accept some of the equations would be to accept a cosmos of such magnitude that it required a leap of imagination they could not make. Proxima Centauri, our closest star apart from the Sun, is almost 25 trillion miles away. The fastest-traveling spaceship built so far would take 18,000 years to get there. Even in the twenty-first century we struggle to understand these distances. The things the Greeks worked out, using what they had, are among the greatest intellectual and scientific achievements in humanity's long history.

As Greek power faded, the Romans had the opportunity to advance the science of astronomy. However, they never embraced mathematics with quite the same passion. The Greeks were interested in astrology, but the Romans were obsessed with it, especially after the founding of the Roman Empire in 27 BCE. Never mind the distance from Earth to the Sun, what was Mars doing in relation to Venus? The life of the emperor might depend on it! The Romans continued to use astrology to make political predictions all the way up until the collapse of the Western Empire in the fifth century, an event they might not have seen coming.

During this period the Chinese had been developing their astronomical skills and finding ways to divide time for practical uses. The mathematician Zu Chongzhi (429–500 CE) devised the "Calendar of Great Brightness" based on 365 days a year over a 391-year cycle, with the need for just one extra month inserted in 144 of the years. Zu wrote that his findings did not derive "from spirits or from ghosts, but from careful observations and accurate mathematical calculations."

Behind Zu's methods was the same ethos that drove the Greeks—the study of empirical facts to explain the world. But the gods and ghosts still dominated thinking in most parts of the world. It would take an explosion of brilliance in the Islamic realm to make great leaps in our understanding.

From the eighth to the fifteenth centuries, across a vast region stretching from what are now the Central Asian Republics to Portugal and Spain, Islamic culture first mastered Greek astronomy and then took it forward during the period known as the Golden Age of Islamic learning. In 900, Al-Battani reduced the length of a year by just a few minutes, and by doing so suggested that Earth's distance from the Sun varied. That in turn suggested that perhaps the planets did not move in perfectly circular orbits. Some scholars began to question the idea that Earth did not move, and it became accepted that it rotates.

A brilliant polymath named Nasir al-Din al-Tusi challenged parts of the Ptolemaic system that were not based on the principle of uniform circular motion. However, again the leap was not made to the model of Earth moving around the Sun.

As Islam's Golden Age blazed bright, Europe was in what used to be called the Dark Ages. Historians now prefer the less pejorative "Early Middle Ages," meaning roughly from the fifth through tenth centuries, from the fall of the Roman Empire to the beginnings of a return to urban life in Europe. It was a time when there was a place for everything, and everything was in its place. All celestial bodies circled Earth, which was the center of the universe. Above this was God; on Earth there were kings, bishops, barons, and serfs; and everyone should be content with their lot. As serfs tended to be unable to write, it isn't easy to know if they agreed. The term "Dark Ages" comes from the Italian scholar Francesco Petrarch (1304–74), who felt that Europeans were living in darkness compared to the brilliance of the Greeks and Romans. In his epic *Africa* he wrote: "This sleep of forgetfulness will not last forever. When the darkness has been dispersed, our descendants can come again in the former pure radiance." Petrarch lived on the cusp of the Renaissance—a time he might well have thought of as "pure radiance." It certainly was for astronomy and its role in progressing humanity's understanding of its place in the universe.

None of the great scientific texts on astronomy were available to Europeans during the Early Middle Ages. This began to change with the work of Gerard of Cremona (1114–87) and others who translated them from Arabic. Gerard went to Toledo to learn Arabic well enough to translate Ptolemy's *Almagest* into Latin (the original Greek edition had been lost for years). It was the first of eighty works he and his scholars translated. The revival of learning was one of the foundations of the Renaissance, opening the doors to knowledge, and the facts flowed in as generation after generation built on what came before and contributed to what is known as the Scientific Revolution, starting in the sixteenth century. It was hard going. The Earth-centered views of cosmology had been adopted by the Catholic Church, and woe betide the heretic who sought to disprove them.

European astronomy took centuries to match the expertise of the ancient Greeks and the Islamic Golden Age. It wasn't until 1543 that it broke serious new ground. That year, the Polish astronomer Nicolaus Copernicus (1473–1543) published *On the Revolutions of the Heavenly Spheres*, which suggested that an Earth-centered universe was wrong.

Copernicus was careful with his phrasing, writing, "if the Earth were in motion." At first criticism was mostly muted. He was a loyal member of the Church and had written "if." He also helpfully died two months after the books came out. However, Catholic and Protestant clergy were keen to undermine his claims, and science was put on notice that the teachings of the Church could not be challenged.

In 1584, the Italian astronomer Giordano Bruno (1548–1600) published *On the Infinite Universe and Worlds*, in which he defended Copernicus and argued that the universe is infinite, with infinite worlds, inhabited by intelligent beings. He was put on trial, and after eight years behind bars he refused to renounce his views, was declared a heretic, and burned at the stake—although it's likely his questioning of more fundamental Catholic doctrine such as transubstantiation played a bigger role in his demise than his views on cosmology.

Next up was Galileo Galilei (1564–1642), the first person to use the newly invented telescope to systematically record observations of the night sky. In 1610, he published *The Starry Messenger*, which made his name and, thanks to its challenge to the idea of an Earth-centered universe, almost cost him his life.

Galileo's studies of the movements of the other planets in the solar system appeared to be in line with Copernicus's theory that Earth did move around the Sun. It wasn't long before Galileo was accused of heresy. The charge was that his beliefs contradicted the Bible—specifically Joshua 10:13, in which a call is made for the Sun to stop moving: "And the sun stood still, and the moon stayed, until the people had avenged themselves upon their enemies." If Scripture said the Sun moved, who was to say it did not?

The pope ordered the theory to be banned. The Church knew that these dangerous new ideas could cause an earthquake undermining the hierarchical model of society, their legitimacy, and ultimately their power. If Earth was not the center of the universe—indeed, if there was no known center—then were humans so important? The French theologian and philosopher Blaise Pascal (1623–62) realized the implications: "Engulfed in the infinite immensity of spaces whereof I know nothing, and which know nothing of me, I am terrified."

Galileo stepped away from the controversy for a while, but in 1623 a new pope, Urban VIII, was elected, who encouraged Galileo to write on the topic, essentially asking him to show his support for the geocentric view. Galileo published *Dialogue Concerning the Two Chief World Systems*, in 1632. It was a nuanced book but came down in favor of the probability that Earth was moving. The pope was not amused, and a two-month-long trial began.

Galileo's defense was that his intent hadn't been to support the Copernican view, that his work was only a means of discussing the view. To no avail—he was found guilty of "having believed and held the doctrine (which is false and contrary to the Holy and Divine Scriptures) . . . that the earth does move, and is not the center of the world." He was sentenced to house arrest, under which he remained until his death in 1642, and was told, "Thou shalt recite once a week the Seven Penitential Psalms."

It could have been worse. Had Galileo not been the most famous scientist in the world he might well have suffered the same painful death as Giordano Bruno. In 1992, 359 years later, the Vatican finally admitted it was wrong.

Despite the wrath of the pope (but probably not God), the tide of knowledge was flowing in the wrong direction for the priests. Our study of the skies had overturned centuries of accepted wisdom and led to a completely new view of the world. The old gods were being challenged—whether that was the intention or not.

A year after Galileo's death, Isaac Newton was born. He went on to invent a new telescope allowing a deeper view into space than had previously been possible. His *Principia* (1687) announced to the world the laws of motion and gravity and ushered in a new age in physics and astronomy.

Newton came not to bury God but to praise him. The more he discovered about the universe, the more convinced he was that its magnificent design must have had a designer: "This most beautiful system of the sun, planets and comets, could only proceed from the counsel and dominion of an intelligent and powerful Being."

Newton agreed that Earth orbited the Sun. Galileo had conducted experiments on what we now call gravity (supposedly dropping objects from the Leaning Tower of Pisa), but Newton's great leap forward was his theory that the laws of gravity applied to all objects, and that this was as true in space as it was on Earth. As with the giants before him, he arrived at a revolutionary moment in history by a combination of empirical work and just sitting down and thinking.

Why did the apple fall in a straight line to the ground? Why did a cannonball fall in a curve as it lost speed? What strange force pulled them down? Newton's law of universal gravitation stated that all objects attract each other, with the force exerted depending on the mass of the objects and the distance between them. So even if the apple was thrown forward from the highest mountain, at such a speed that it just kept going, it would not head out into space in a straight line but "fall" around the world in a never-ending curve, held close to Earth by this strange force called gravity, from the Latin *gravitas*, meaning "weight." And gravity, he said, explained why planets constantly revolved around the Sun instead of just wandering off into space. The closer a greater object is to a smaller one, the stronger is its gravitational pull.

There was some limited resistance to his ideas by a few scientists on the grounds that Newton's gravity was akin to primitive superstitions about a supernatural power. He was content to prove his ideas rationally, and to believe in his God.

There was more, so much more. Newton's work is considered by some to have been the greatest contribution to the history of science. When he died in 1727, his body lay in state in Westminster Abbey for a week. The great English poet Alexander Pope wrote, "God said, *Let Newton be!* and all was light."

This was an exciting time for science, akin to that of the ancient Greeks and the Islamic world's Golden Age, but different in that knowledge advanced more quickly than before. Each discovery created another chink in the armor of organized religion and its claim to power. In the Age of Reason, it became unreasonable to tell a scientist to recite the Penitential Psalms for contradicting Scripture.

Staring up at the sky had led to a complete revolution in the way we thought and lived our lives, opening up the road to further scientific endeavor. Gradually, but not entirely, organized religion in the technologically advanced countries retreated to its temples, and science occupied the temporal sphere.

It was an age of miracles and wonders. Since then we have learned a huge amount more, and there's a majesty to our science that now allows us to see so much when we gaze up at the stars. A modern space telescope can look back in time and detect light that has been traveling for more than 13 billion years before hitting the lens.

In 1931, Georges Lemaître suggested that the universe began with the explosion of a single tiny particle, which he called the "primeval atom." This idea was supported by evidence in the 1920s from Edwin Hubble's massive telescope on Mount Wilson in California, which appeared to show that all observable galaxies were moving away from Earth in every direction at rapid speeds. It was logical to conclude from this that they must have originated from a single place at a specific point in time. This theory would become known as the Big Bang. At the time, conventional wisdom mostly supported the Steady State theory—that the universe had always existed and always would. But in the 1950s, new measurements of the speed of movement of the galaxies suggested its birthday was 13.7 billion years ago. This was an extraordinary revolution in our understanding of the universe.

In April 1990, the twelve-ton Hubble Space Telescope was put into orbit. Free from the limiting and distorting effects of Earth's atmosphere, the telescope began to bring the cosmos into sharper focus and to look further and further into its past, to within microseconds of its, and our, birth. Now, infrared telescopes can detect light from radiation that can pass through cosmic dust but cannot be seen by the human eye or visible light telescopes such as the Hubble. Measuring the wavelengths and composition gives the data to tell the story of the universe.

All these discoveries have been driven by the need to answer the questions "How?" and "Why?" Science is brilliant at answering the first, but even when it finds the answer, it often throws up yet another "Why?" Despite our advancing knowledge, we have still not taken the wonder out of the universe. In many ways, the theories and discoveries of the twentieth century only added to it, posing questions that might only be answered as we begin to explore the physical realities of space.

During the first two decades of the last century the world was introduced to the strangeness of quantum mechanics and Albert Einstein's theories of relativity and space-time. Quantum theory suggests that the mysterious subatomic world of tiny particles is governed by total randomness, an idea that conflicted with Einstein's (and Newton's) view that there are universal laws. The debate is worth touching on briefly. Briefly because most of us are in good company with some of the best brains ever to have existed in that we, and they, don't really understand quantum mechanics. Nevertheless, it, Einstein's response, and his discoveries tell us something about why our destiny is in space.

Quantum entanglement theory suggests particles can instantly influence one another even if they are hundreds of millions of miles

25

away. The key word here is "instantly." But this simply doesn't fit with the accepted idea that there are universal laws of science. For example, as Einstein showed, nothing can travel faster than the speed of light.

That is why he rejected quantum entanglement as "spooky action at a distance" and scientists continue to argue about its validity. Nevertheless, it leaves open the possibility that laws are not universal. If so, perhaps something can travel faster than the speed of light, implausible as this sounds. One of Einstein's most famous quotes was in response to this dilemma: "God does not play dice with the universe."

Einstein agreed with Newton that space has three dimensions: height, width, and length. But Newton thought that the objects in space did not affect these dimensions. Einstein said they did. His Theory of Special Relativity had added a fourth dimension, time, and he called this combination of four dimensions space-time. This fourth dimension could be warped by large masses even to the extent of speeding it up or slowing it down. Think of space as a foam mattress. You step on it. Your weight (or mass) causes a depression in space. According to Einstein, gravity is a distortion in the shape of space-time.

Our ancestors looked up and saw a universe they could not understand but used its apparent order to make sense of their world. We now know so much more, and yet still confront an infinite universe full of mystery containing dark matter, black holes, warps in the fabric of space-time, and challenges to the very concept of order and law. This is what Newton meant when he said, "What we know is a drop, what we don't know is an ocean."

The implications of quantum mechanics and space-time on what will, and will not, be possible in space travel are unknown but will potentially open new avenues in the distant future. Because after all these millennia of discovery, there are still more questions than answers, and more questions to be asked that we don't even know yet. Some of those questions and answers will only be found the farther from Earth we go. And the desire to find out, to know more—and even go there ourselves—has proved irresistible.

Astronaut Edwin Aldrin on the Moon beside the US flag, July 21, 1969.

2

THE ROAD TO THE HEAVENS

I see Earth! It is so beautiful!
—Yuri Gagarin

We first crossed the border with space less than a century ago. It had taken thousands of years of slow development, followed by an amazing sprint during those decades of miracles and wonders in the twentieth century. But it was conflict on Earth that finally got us there. The technology that took us to the heavens came from the arms race of the Cold War.

For almost all of human history it was so near and yet so far. As the British astronomer Fred Hoyle said in 1979, "Space isn't remote at all. It's only an hour's drive away if your car could go straight upwards." Formula One engineers can soup up their car engines as much as they like, but they won't top out at the 4.9-miles-a-second speed required to leave Earth's surface and go into orbit. A rocket engine, on the other hand . . .

Such a simple thing, a rocket. So simple that we can buy them in shops and launch them from our back gardens to celebrate birthdays or New Year's Eve. Conversely, getting one into space with a human being in it is so fiendishly complicated that only three countries have done it.

One of the difficulties of human space travel is that the cutting-edge technology required ultimately relies on putting people on top of giant tanks of fuel. Then setting fire to the fuel. Space Shuttle astronaut Mike Massimino best captured the spirit of this in his memoir, *Spaceman*. He wrote about looking at his cheerful colleagues as they approached the launchpad: "Are they insane? Don't they see we're about to strap ourselves to a bomb that's going to blow us hundreds of miles into the sky?"

Indeed. The shuttle's external fuel tank held 172,000 gallons of liquid oxygen and 449,092 gallons of liquid hydrogen. The engines then burned this at the rate equivalent to emptying a family swimming pool every ten seconds.

This basic technology is not so different from that discovered by monks in China in the ninth century using gunpowder: a mix of sulfur, potassium nitrate, and charcoal. At first it was used for fireworks, but the Chinese moved on to make "flying fire lances." In the sixteenth century, one man even supposedly tried to use these to reach the stars. As the Chinese legend tells it, Wan Hu attached forty-seven gunpowder-filled rockets to a bamboo chair, strapped himself to it, and ordered his servants to light the blue touchpaper. He then traveled a short distance upward before disappearing in a massive explosion and clouds of smoke. He was never seen again, nor was the chair. There isn't any written evidence that the event happened. However, there is now a crater on the Moon named after Wan Hu.

Over the centuries there have been other attempts to design rockets, with varying degrees of success, but when it comes to the lineage of modern rockets, historians of spaceflight usually reference three names: Konstantin Tsiolkovsky (1857–1935), Robert Goddard (1882–1945), and Hermann Oberth (1894–1989). All were brilliant pioneers in their field. Goddard, an American, was the first person to get a rocket off the ground using liquid fuel rather than the compressed powder of solid fuel that had been used since the Chinese discoveries of the ninth century. Oberth was a German scientist whose reputation is tarnished by having worked for the Nazis. They used his studies on rockets to develop the Vergeltungswaffe 2 (Vengeance Weapon 2), or V-2, rocket that was used to such devastating effect against civilian targets during the Second World War. He also conducted medical experiments on himself to support his theory that humans could survive the physical stresses of space travel, such as g-force and weightlessness. Arguably, though, the most impressive of the three, for sheer brilliance of imagination, is Tsiolkovsky.

In 1903, seven months before the first powered aircraft had flown, an unknown, self-taught Russian scientist published the first theoretical proof for the possibility of spaceflight. Later that year the Wright brothers flew into the history books, but Tsiolkovsky remains virtually unknown, despite being one of the most farsighted scientists to have lived.

Born the fifth of eighteen children to parents of modest means, he became deaf at age ten after a childhood illness, left school at fourteen, and went on to learn science from reading books in a public library, including numerous volumes on physics, astronomy, and analytical mechanics, as well as the science fiction of Jules Verne. "Besides books I had no other teachers," he wrote.

His early writings included visionary ideas: how to build space stations powered by solar energy, sketches of gyroscopes to control a spaceship's orientation, air locks to enable spaceships to dock with each other, and pressurized space suits that would allow cosmonauts to venture outside their craft. As early as 1895, he was theorizing the concept of a space elevator. He went on to produce a stunning body of work including the 1903 paper that later propelled him to fame in Russia. "Exploration of the World Space with Reaction Machines" contained the first scientific theoretical proof that a rocket could push through the atmosphere and orbit Earth. Tsiolkovsky had worked out the horizontal speed required to get into orbit, and that it could be achieved using rockets fueled by a mixture of liquid hydrogen and liquid oxygen. His formula, known as "the Tsiolkovsky rocket equation," set out the relationship between the speed of the rocket, the changing mass of the rocket and its fuel, and the speed of the gas as it is expelled. It is the foundation of space travel.

When the Soviets took over, they were suspicious of Tsiolkovsky's quasi-theological musings on space travel, which were at odds with Communist philosophy. In "Is There God?" he argued: "We are at the will of and controlled by Cosmos . . . we are marionettes, mechanical puppets." In fact, he was controlled by the Communist Party. At one point the secret police arrested him, and he spent several weeks in the notorious Lubyanka jail in Moscow accused of anti-Soviet propaganda.

However, as the fledgling rocket industry got off the ground, the Soviets realized the PR benefits of claiming a pioneer as one of their own and in 1929, Tsiolkovsky was allowed to publish the first paper proposing the concept of a multistage rocket booster.

The prophet is not without honor, especially in the land of his birth where he has many epitaphs, from "father of spaceflight" to "father of rocketry." His modest log cabin is open to the public; nearby stands the State Museum of the History of Cosmonautics, which bears his name. On the far side of the Moon, a huge crater discovered by the Soviet spacecraft Luna 3 is named after the man who knew that science fiction can become science fact.

Knowledgeable science fiction experts know all this. In the comic book series *Assassin's Creed: The Fall*, a lead character reads from Tsiolkovsky's *The Will of the Universe*. In an episode of *Star Trek: The Next Generation* a space ship is named after him. He is quoted in two of Sid Meier's video games, and has been name-checked in a short story by the sci-fi writer William Gibson. Meier and Gibson would undoubtedly know Tsiolkovsky's most famous quote: "Earth is the cradle of humanity, but one cannot stay in the cradle forever." Shortly before his death he wrote: "All my life I have dreamed that by my work mankind would at least be advanced a little." It was.

———

Putting theory into fact wasn't easy. To achieve Tsiolkovsky's equation you must accelerate. To accelerate you need fuel. The faster you accelerate, the more fuel you need. The more fuel you need, the heavier the craft carrying it becomes.

In the first few decades of the twentieth century, many scientists were grappling with this problem. The decades prior to the Second World War saw various advances, but it was the war itself, and then the Cold War, that led to rapid advances in technology, driven by the desire to win.

The Soviets and Japanese both experimented with rocket-powered planes and Japan even developed a rocket-powered kamikaze bomber. But it was the German rocket program that led the way. Overseeing it was Wernher von Braun, a Prussian aristocrat who was inspired by the work of Hermann Oberth. As had Oberth, von Braun joined the Nazi Party and became a major in the SS.

In 1942, he oversaw the first launch of a rocket into suborbital space, about 62 miles up, but his team could not yet engineer a rocket that was able to achieve the speed required to enter orbit. However, his V-2 could travel at more than 300 miles per hour, and for about 200 miles, before falling back to Earth. When Adolf Hitler was told about von Braun's breakthrough, he tasked him with building thousands of them, tipped with warheads. In 1944, the first V-2s were launched. Traveling faster than the speed of sound, they were almost impossible to intercept and hit their targets less than three minutes after being launched.

As Hitler's "Thousand-Year Reich" began to implode twelve years after its inception, von Braun and his team headed to Bavaria and surrendered to the Americans. It was a good move, given that the alternative was to surrender to the Russians. Both powers had intelligence officers tasked with finding both the Nazi secret weapons and the scientists who made them.

In what became known as Operation Paperclip, von Braun and about 120 other German scientists were secretly flown to the US to develop America's ballistic missile program. The scientists' pasts were covered up. Many were ardent Nazis but, unlike some of their counterparts who faced justice at the Nuremberg trials, instead of being hanged they were hired. The V-2 rockets had been built mostly by slave laborers handpicked from Buchenwald concentration camp by von Braun himself, and they killed thousands of civilians.

The cheerful and articulate von Braun eventually became the director of NASA's Marshall Space Flight Center and the public face of the American space program. He was said to have remarked about his V-2 rockets that they had worked perfectly, except for landing on the wrong planet. His moral detachment was matched by that of the Americans, who made a Faustian pact with him, whitewashing his past in order to help them fight the new war they found themselves in—the cold one.

The Russians took a similar view. Their version of Paperclip was Operation Osoaviakhim. In October 1946, Soviet army and intelligence units took more than 2,200 German scientists and their families to Russia to work on various projects, including the rocket program. The Cold War had begun.

It was a time when people across the world lived in the shadow of the mushroom cloud. Children practiced "duck and cover" drills to survive a nuclear attack and people were encouraged to build their own air-raid shelters even though they would be of no help in the event of a thermonuclear exchange. In August 1949, the Soviet Union detonated its first atomic bomb at a remote test site in Kazakhstan. A US spy plane flying off the coast of Siberia picked up traces of the radiation, and a few weeks later President Harry Truman announced to the world that the Soviet Union was a nuclear power. Nuclear war between the two countries was now a possibility. The dangers of a nuclear holocaust only grew when both developed hydrogen bombs, even more powerful than the atomic versions.

Among the weapons used in the Cold War was technology, deployed by each side to prove that its political system—and armory—was superior. By the 1950s, they were building ballistic missiles that could launch satellites into space to test density levels in the atmosphere, study radio wave transmissions, and track objects in orbit. Of course, the missiles had another purpose as well.

The Soviet space program was headed by Sergei Korolev. In the 1930s, under torture, he'd "confessed" to being a counterrevolutionary against the motherland and was sent to a notoriously brutal gulag in Siberia. There he was starved, had his teeth knocked out, and his jaw broken, but as war with Germany loomed, he was transferred to a Moscow jail, where he worked on rocket designs during the war. During the Cold War his orders were: "Beat the Americans, get there first." He made it with four months to spare.

In early October 1957, several amateur radio enthusiasts in the eastern United States picked up a series of *beep-beep-beep* sounds on their short-wave radios. Some recorded them and within hours the American television and radio audiences were listening to transmissions from Sputnik 1—the first man-made object to orbit Earth. The threshold had been crossed. The Space Age was under way.

Sputnik 1 was launched on October 4 from Kazakhstan. It was barely bigger than a beach ball, weighing a mere 184 pounds. It had four long antennae protruding from its sphere and inside were a thermometer, a few batteries, a radio transmitter, and a fan to keep it cool. The Americans got very hot under the collar.

It was hailed as a victory for Russia, the Soviet Union, and communism. The newspaper *Pravda* commented: "All the world heard the announcement of the launching of the artificial moon." The Soviet leader Nikita Khrushchev learned of its success at 11 p.m. at a drinks reception in the Mariinsky Palace in Kyiv. His son Sergei recalled that Khrushchev was told he had a call and left the room, returning a few minutes later, "his face shining." He then sat silently for a time before raising his hand for silence. "Comrades," he told senior members of the uncomprehending Ukrainian Party leadership, "a little while ago, an artificial satellite of the Earth was launched."

The White House pretended not to care. President Eisenhower called it a "small ball in the air," an aide said the US was not playing "an outer space basketball game," and another even called Sputnik "a silly bauble." In private, though, the significance of Moscow's achievement was sinking in, and the US headlines made anyone doubting the enormity of the event concentrate—"A Grave Defeat" declared the *New York Herald Tribune*, a "National Emergency" said *The Reporter*. The small ball in the air had shattered the US's sense of invulnerability.

Sputnik 1 had a highly polished aluminium exterior that shone so brightly Americans could see it as it passed overhead every ninety minutes, every day, for three months, before it burned up after reentering Earth's atmosphere. Every time it came past was another reminder that the Soviets had surpassed American technology. The anxiety in the United States was not so much about the satellite as about the massive rocket that had carried it into space. What the Russians called Iskustveni Sputnik Zemli, or Artificial Satellite of Earth, was a game changer. Before Sputnik, the US assumed it would be able to intercept Soviet nuclear-armed aircraft. But Sputnik had been delivered to space on top of what was in effect a ballistic missile, which it was now clear could reach America.

The historian Walter McDougall later spoke of the effect the news of Sputnik had on the American government and people: "To have the communists lead in technology? To pioneer a new frontier of infinite size? In a sense to capture the future? . . . What did this mean? That the future belongs to communism?" Now the Reds weren't just under the bed—they were overhead.

A memo marked "Confidential" written for the White House a few days after Sputnik launched gives an insight into what Eisenhower's administration thought was at stake. Titled "Reaction to the Soviet Satellite," it says: "Public opinion in friendly countries shows decided concerns over the possibility that the balance of military power has shifted," and ends: "General Soviet credibility has been sharply enhanced." A few weeks later the Soviets successfully launched Sputnik 2. Inside was a dog named Laika who became the first animal in space, but sadly not the first to return.

Eisenhower gave the go-ahead for an American satellite to be launched as soon as possible. Two months after Sputnik 1 lifted off into space, the rocket carrying America's Vanguard Test Vehicle Three (TV3) set off from Cape Canaveral, rose just over three feet, fell back to Earth, and exploded. In contrast to what had happened in the USSR, the news cameras had been invited to record the event and the outcome was broadcast coast-to-coast within hours. The media had a field day with headlines such as "Kaputnik!" and "Flopnik." The Soviets offered the US help under their "programme of technical assistance to backward nations."

Eisenhower was not amused. The US's budget for its space program jumped from 0.1 percent of the federal budget in 1958 to 0.5 percent in just two years, and under the Kennedy and then Johnson administrations it grew to 4.41 percent. In January 1958, the von Braun–designed Juno 1 rocket successfully took America's Explorer 1 satellite into orbit. But the Soviets had achieved two "firsts." Both sides now looked for the next.

Over the years that followed, each had a few, but none was of the magnitude of Sputnik 1. In December 1958, President Eisenhower's recorded Christmas message was transmitted from a US satellite and became the first broadcast of a human voice from space. A couple of weeks later the USSR's Luna 1 probe missed its intended target of the Moon, sailing right past it, and began to orbit the Sun instead of Earth—a first, but an accidental one. It's still out there—circling the Sun once every 450 days.

Then, later in 1959, the Soviets had a hit, literally, when Luna 2 became the first spacecraft to reach the surface of the Moon. It was a "hard landing," which is scientific talk for "crashed," but it did its job and scattered silver panels bearing Soviet symbols on the surface. In a nice touch Khrushchev sent a replica of one to President Eisenhower as a gift. That year also saw Luna 3 (another Korolev design) reach the far side of the Moon. It was, as it often is, bathed in sunlight, but years later Pink Floyd was not going to let that stand in the way of their best-selling LP.

The year 1960 saw the Americans launch a Television Infrared Observation Satellite (TIROS) to study the weather. Within days it was able to detect and track a storm off the coast of Madagascar and eventually became the prototype of the current global systems used for weather reporting. It could capture only large-scale features, but that was still enough to make Moscow nervous. Later that year Sputnik 5 flew two dogs, Belka and Strelka, to space and, happily for them, returned them alive. After a period as a celebrity, Strelka retired from public life and had six puppies, one of them named Pushinka (Fluffy). Khrushchev remembered that during a conversation in 1961 with First Lady Jacqueline Kennedy she had asked after Strelka. Now developing a skill for gifting, he sent Pushinka to the White House, complete with a Soviet passport. President John F. Kennedy wrote to thank him: "Mrs. Kennedy and I were particularly pleased to receive 'Pushinka.' Her flight from the Soviet Union to the United States was not as dramatic as the flight of her mother, nevertheless, it was a long voyage and she stood it well. We both appreciate your remembering these matters in your busy life." Pushinka and one of the Kennedy dogs, Charlie, then developed a shine for each other, resulting in four puppies referred to by JFK as "pupniks." Given the extreme tensions of the Cold War, these rare moments of cordiality were welcomed.

But there was still a Space Race to win. The Americans saw Belka and Strelka and raised them Ham—a chimpanzee who became the first hominid sent into space on January 31, 1961. No one remembers Ham, though, because the second hominid sent into space was also the first man in space. The Americans had unfortunately named their project "Man In Space Soonest," or MISS. They did.

On April 12, 1961, Senior Lieutenant Yuri Alekseyevich Gagarin approached the Vostok 1 rocket, pausing only to urinate on the right-hand back wheel of the vehicle that had taken him to the launchpad. To this day, Russian cosmonauts do the same in tribute to him. (Female crew members splash the wheels with liquid from a bottle.) Gagarin then climbed aboard the capsule and waited. There was no countdown—Sergei Korolev thought of them as an American affectation—and at 9:07 a.m. Moscow time they simply pressed a button. Gagarin shouted *"Poyekhali!"* ("Let's go!"), and off he went, slipping the surly bonds of Earth into what the poet and pilot John Gillespie Magee Jr. had called "the high untrespassed sanctity of space" and engraving his name in the annals of the human story.

The flight lasted 108 minutes as Gagarin made just over one orbit of Earth. Upon reentry, about 4 miles up, he ejected from the capsule and landed in a rural area of the Volga region. A few minutes later, a woman named Anna Takhtarova and her five-year-old granddaughter saw a spaceman wearing a bright orange suit and white helmet walking toward them across a field where they had been planting potatoes. Gagarin later recalled, "When they saw me in my space suit and the parachute dragging alongside as I walked, they started to back away in fear. I told them, don't be afraid, I am a Soviet citizen, like you, who has descended from space and I must find a telephone to call Moscow!"

Gagarin became a global celebrity, a "Hero of the Soviet Union," and a major asset to the communists in the Cold War. He was only twenty-seven, charming, and had a ready smile. Better still, he was also the son of peasants from a small collective farm and had risen to become a fighter pilot, then a cosmonaut, and then the first man in space—what better proof of the superiority of the Soviet system over the capitalist West?

Gagarin was chosen from among two hundred fighter pilots enrolled in the Soviet program. Ahead of the launch they had been whittled down to two. His rival was Gherman Titov, every bit as able as Gagarin but with a flaw—he came from a comfortable middle-class, well-educated family. Khrushchev knew the propaganda value of the "from collective farm to space" narrative, and so the peasants' son rode Vostok 1 through the atmosphere and into space. Before attending his victory parade in Red Square, Gagarin's parents were told to wear simple clothes at the event.

The story broke in the United States in the early hours and news desks across the country began to call NASA for comment. Duty officer John "Shorty" Powers, cross that his slumbers had been disturbed, shouted at one reporter, "What is this! We're all asleep down here!," resulting in the classic headline: "Soviets put man in space. Spokesman says US asleep."

It was quite the wake-up call. A few months earlier, in his inaugural address, President Kennedy had said, "We shall pay any price, bear any burden, meet any hardship, support any friend, oppose any foe to ensure the survival and the success of liberty." After Sputnik and Gagarin, massive funding for NASA was part of that price.

On May 5, 1961, just three weeks after Gagarin landed, Alan Shepard became the first American, but second man, to travel to space. Kennedy set his country's sights higher. He and Vice President Lyndon Johnson had concluded that orbiting the Moon or building a space station would not be enough to demonstrate American technological prowess and leadership. For that they had to land Americans on the Moon and show the world they'd done it. He laid it out in a speech to Congress that same month, saying: "If we are to go only halfway, or reduce our sights in the face of difficulty, in my judgment it would be better not to go at all."

He also made clear the connection with the Cold War: "If we are to win the battle that is now going on around the world between freedom and tyranny, the dramatic achievements in space which occurred in recent weeks should have made clear to us all, as did the Sputnik in 1957, the impact of this adventure on the minds of men everywhere . . . I believe that this nation should commit itself to achieving the goal, before this decade is out, of landing a man on the moon and returning him safely to the earth. . . . It will not be one man going to the moon—if we make this judgment affirmatively, it will be an entire nation."

The spirit of the times was caught the following year in his "We choose to go to the Moon" speech in Houston: "We choose to go to the moon in this decade and do the other things, not because they are easy, but because they are hard." Von Braun set to work.

Korolev was already busy. Despite his many successes, including Sputnik 1, his role as chief designer of the Soviet rocket program was unknown to the public. It was only revealed after his death in 1966 following complications during routine surgery. Doctors tried to use a breathing tube but could not get it down his throat because it had been so damaged in the gulag. Korolev was given a state funeral and his ashes taken to the Kremlin Wall. Gagarin read the eulogy.

Two years later, he too was gone. He'd said about his journey to space, "I could have gone on flying through space forever," but it was flight that killed him while test-piloting a MiG-15 fighter jet at age thirty-four. Tens of thousands of people attended his funeral in Red Square and his ashes were interred close to Korolev's.

Between Kennedy's speeches and Korolev's death, the Soviets had kept up their string of "firsts," all of which had the Russian engineer's stamp on them. First dual-crewed spaceflight, 1962. First woman in space, Valentina Tereshkova, 1963. First spacewalk, Alexei Leonov, 1965. Leonov's spacewalk was dramatic enough, but while outside his craft, Leonov's suit swelled up, making it impossible for him to get back into the capsule. There were several tense minutes as he bled enough oxygen out to allow him to squeeze back through the three-foot-wide air lock. A year later, Luna 9 achieved the first soft landing on the Moon and transmitted the first close-up photos of its surface.

In response to Kennedy's 1961 speech Khrushchev had refused to confirm or deny that Moscow was in a race to the Moon. Secretly he had given the order: If the Americans said they would be on the Moon "before this decade was out," the Soviets would be there before them, aiming for 1968. Not without their chief designer and chief inspiration Sergei Korolev they wouldn't.

Following his passing there was a series of technical failures, including the tragic death of Vladimir Komarov, the pilot of Soyuz 1, in 1967. After several mishaps his mission was aborted, only for the craft's primary parachute to fail and the reserve chute to become entangled. Soyuz 1 hit the ground at high speed and exploded. It took engineers eighteen months to find and fix the problems before piloted missions could fly again. NASA had its own tragedies, including the deaths of Virgil "Gus" Grissom, Ed White, and Roger Chaffee in a fire in the Apollo 1 cabin during a ground test in 1967. It took almost two years before the faults identified could be rectified.

But the race for the first crewed Moon landing was still on. The Soviets were aware of the difficulties NASA was having with the Saturn V rocket it had developed for launch, and the lunar landing vehicle, and concluded that the US would miss its deadline and would not try until 1970 at the earliest. Many in NASA felt the same way. Conversely, the Americans, unaware of the scale of problems the Soviets were facing post-Korolev, feared they would use a launch window that was coming up in December 1968, after which the Moon would not be in a proper position for flights until well into 1969.

The window opened, then closed, with no movement from the Soviet side. But in the same month three Americans became the first men to orbit the Moon. Apollo 8 circled it ten times with Frank Borman, Jim Lovell, and Bill Anders on board. Anders took the famous *Earthrise* photograph and said later they'd gone to the Moon but discovered Earth. The image of our planet hanging precariously in the void, with its thin atmospheric layer protecting it, had a huge psychological effect on many people who saw it and is credited with giving a great boost to the fledgling environmentalist movement. On Christmas Eve, before they returned home, all three participated in a live TV transmission and took turns reading from the book of Genesis:

And God said, Let there be light: and there was light.

And God saw the light, that it was good: and God divided the light from the darkness (KJV).

Numerous sources put the global viewing figures at a billion people—about one in four humans. That seems unfeasibly high, but it was without doubt a massive audience for an amazing event. Humans had been around the Moon and back. Next up was the main aim. The clock was ticking.

"T minus ten, nine, eight, seven . . ." It was July 16, 1969. The countdown for Apollo 11 was under way. Korolev had been right—the countdown was an American affectation. Or rather, an American-German affectation. The 1929 Fritz Lang film *Frau im Mond* (*Woman in the Moon*) had featured the first rocket launch countdown to heighten the tension and used captions reading "*Noch 10 sekunden*" ("10 more seconds") etc., culminating in *"Jetzt!"* ("Now!") Guess who saw the film . . . a young Wernher von Braun, who was much taken by the idea. It married well with the American sense of drama and spectacle in the television age.

It doesn't get much more dramatic than a crewed rocket launch, and it's worth revisiting Space Shuttle astronaut Mike Massimino's memoir for a fraction of an insight into what astronauts Neil Armstrong, Edwin "Buzz" Aldrin, and Michael Collins went through at the Kennedy Space Center Launch Complex 39A:

At six seconds you feel the rumble of the main engines lighting. The whole stack lurches forward for a moment. Then at zero it tilts back upright again and that's when the solid rocket boosters light and that's when you *go*. There's no question that you're moving. It's not like *Oh, did we leave yet?* No. It's *bang!* and you're gone. . . . I felt like some giant science fiction monster had reached down and grabbed me by the chest and was hurling me up and up . . . The whole thing can be summed up as controlled violence, the greatest display of power and speed ever created by humans.

Saturn V was the most powerful launch vehicle ever built. It had three stages. The first fired its engines and lifted the 364-foot-tall rocket off the ground while burning 20 tons of fuel per second. Before it had even cleared the launch tower it was traveling over 62 mph. After two and a half minutes, and 42 miles up, the first stage ran out of fuel, fell away, and the second stage ignited its engines. Six minutes later, Saturn V was up well over 100 miles and accelerating toward orbital velocity. As the second stage fell away the third took over, sending Armstrong, Collins, and Aldrin into orbit at 17,398 mph.

The rest of the journey took just over three days. On the way they checked that they were on course using an instrument familiar to Galileo—a telescope—and another known to generations of sailors, a sextant. The computer on board the command module was less powerful than a modern pocket calculator. It was a tense descent as Armstrong and Aldrin brought the Eagle lunar module down onto the boulder-strewn surface of the Moon—as they landed it had just fifteen seconds' worth of fuel left in the tank. Four hours later, Armstrong made his small step onto the surface of the Sea of Tranquility and giant leap into history.

July 21, 1969: a date that will be remembered in the distant future as one of the most incredible moments in humanity's story, long after details of many wars, revolutions, stock exchange crashes, and pandemics have faded into obscurity. Armstrong is a colossal figure, but he knew he stood on the shoulders of giants such as Gagarin and Tsiolkovsky, Goddard, Oberth, Korolev, von Braun, and, before them, the great scientists down the ages. He also understood the significance of the moment in the Cold War, saying later: "I was certainly aware that this was a culmination of the work of 300,000 or 400,000 people over a decade and that the nation's hopes and outward appearance largely rested on how the results came out." Among them were unsung heroes such as the brilliant mathematician Katherine Johnson, who calculated the precise trajectories allowing Apollo 11 to land on the Moon, and Margaret Hamilton, who coined the phrase "software engineering" and wrote the programs controlling the command and lunar modules.

Armstrong also knew he was not alone in another sense—the Soviets were overhead. In a last-ditch effort to at least get a machine to the Moon's surface and back, they'd launched an unmanned craft a few days before Apollo 11 took off.

They'd known for months that their dream of being first to land a human on the Moon had almost certainly gone. Or gone up in flames, to be more precise. They were well behind the Americans even before two catastrophic events that year involving the giant N1 rocket, their rival to the Saturn V. The first, in February 1969, saw the rocket and unmanned landing module lift off from the Baikonur Cosmodrome launch center in Soviet Kazakhstan, streak upward for about two minutes, reach an altitude of 8.7 miles before slowing, and then fall back to Earth some way from the launch site, exploding on impact.

In early July, just two weeks before Apollo 11's launch date, the Soviets tried again. Mid-ranking officials had tried to warn the top brass about a series of potential problems but were told to keep quiet. The Politburo in Moscow was told what its senior members wanted to hear. This time the rocket and module only got 328 feet off the ground before appearing to freeze in midair, then tilting over, crashing back down, and exploding. Most of the launch complex was destroyed, and windows in the technicians' residential area 22 miles away were blown in.

Even if the Apollo 11 mission had failed, the USSR would not have had an advantage. It would take more than a year to rebuild the N1 launchpad. But they still had the Proton-K rocket and a Luna module capable of landing on, and lifting off from, the Moon. They could fit it with telecommunications systems, a drilling kit to collect lunar soil, and a camera, and they could launch it and get it back before Apollo 11. A first home run may not be as good as first man on the Moon, but it might dilute the effect of what the Americans were about to do.

Thus, three days before Apollo 11 took off from Cape Kennedy, Luna 15 set off from Baikonur. The Americans didn't know what the launch was for, but the Soviets knew the race was on. The Soviet craft endured technical problems en route and then lost more time as it orbited the Moon, and technicians realized its landing trajectory might take it into rugged terrain in which it would crash. Twice they delayed the landing procedure, and into the gap flew Apollo 11.

By the time the Soviet scientists were confident enough to land Luna 15, Armstrong and Aldrin had been out for their moonwalk, gathered 49 pounds of soil and rocks, planted the American flag, spoken with President Richard Nixon in front of a global TV audience estimated at more than 650 million people, and were back in the spacecraft. Two hours before Apollo 11 took off from the Moon, Luna 15, now on its fifty-second orbit, began to descend.

As the dramatic events were unfolding, British scientists at the Jodrell Bank Observatory were listening in to the transmissions from both missions via a radio telescope. Rumors from Moscow suggested Luna 15 might be equipped to land, and on the recordings made at Jodrell you can hear the moment its mission became clear. In a wonderfully British manner one of the scientists exclaims, "It's landing! . . . I say, this has really been drama of the highest order."

But it was more crashing than landing. It came in at an angle. Data suggests that when its last transmission came, Luna 15 was about 2 miles above the Moon's surface. It probably crashed into the side of a mountain at about 298 mph. The crash site was in the Sea of Crises. Shortly afterward Armstrong and Aldrin took off, leaving behind a commemorative medallion bearing Gagarin's name and those of other cosmonauts and astronauts who had lost their lives in the Space Race.

Exactly 2,982 days had passed since Kennedy had given the deadline for success. They'd made it with 161 days to spare.

The contest was over. The Americans had won, so the Soviets pretended it had been a one-horse race. The USSR, champion of the world's workers, would never have wasted the people's money on such a costly, dangerous sideshow, sniffed the Kremlin. Radio Moscow's message to its Marxist-Leninist allies in countries such as the People's Republic of Angola, the Republic of Cuba, and the Democratic Republic of Vietnam was that Apollo 11 was part of "the fanatical squandering of wealth looted from the oppressed peoples of the developing world."

Despite evidence to the contrary, the lie was believed in some Western circles. As early as 1964, the *New York Times* said, "There is still time to call off what has become a one-nation race," and in 1974, Walter Cronkite told CBS viewers, "It turned out that the Russians were never in the race at all." Similar views were held until 1989 and the Soviet period of glasnost, or openness. Then a team of American aerospace engineers was invited to the Moscow Aviation Institute and shown the lunar landing craft the Soviets had built to get their cosmonauts to the Moon first. The *New York Times* ran a front-page headline: "Now, Soviets Acknowledge a Moon Race." The article read, "There is still time to call off what has become a one-nation race."

After 1969, the Soviets slowly concluded that coming in second wasn't worth the huge sums of money they were spending. The cosmonaut training program was scrapped but the rocket engineers were kept on. A lunar landing in the 1970s would only confirm that they had been trying all along, and that their technology was inferior. As Yaroslav Golovanov, a journalist for *Pravda*, later noted, "Secrecy was necessary so that no one would overtake us. But later, when they did overtake us, we had to maintain secrecy so that no one knew that we had been overtaken."

The Americans went on to complete six crewed missions, and in total landed twelve astronauts on the Moon's surface. Apollo 17 was the last, returning on December 14, 1972, and since then no one has been back. The space program had drained $30 billion from the country's coffers, the Vietnam War was raging, there were riots in the big cities, and public interest in the landings had waned.

The American and Soviet leaders (Nixon and Brezhnev) cut the space budgets, and during a slight thaw in the Cold War the two nations planned a joint mission to dock a Soyuz craft with an Apollo. They came together in 1975 and the two crews exchanged gifts as they visited each other's spaceships via an air lock not dissimilar to the one Tsiolkovsky had designed at the beginning of the century. Both countries then focused on space shuttles and orbital space stations.

And the Moon? It's still there, of course. Also still there are the three vehicles (Moon buggies) the Americans left behind, as well as tools and television equipment abandoned to make room for the soil and rock samples brought home. One day, perhaps, they will be in a museum on the Moon, along with many of the other objects littering the surface. There are several US flags, and a plaque from the Apollo 11 mission that reads: "Here men from the planet Earth first set foot upon the Moon July 1969, A.D. We came in peace for all mankind."

There are also a hammer and a feather. Apollo 15 astronaut David Scott paid tribute to Galileo's experiments in the sixteenth century, when the Italian is said to have dropped two objects of different weights from the Leaning Tower of Pisa. Scott said Galileo was instrumental to the Moon landings. As he dropped a feather and a hammer onto the lunar surface, a television audience watched as they fell at the same speed. The feather came from Baggin, the Air Force Academy's falcon mascot.

And there are two golf balls. Alan Shepard took the head of a golf club onto the Apollo 14 mission, attached it to one of the tools, and hacked his way into history. All these items speak of the romance of space exploration, less so the hundred or so bags of urine and excrement left behind. There may be room for one or two in our future Moon museum, but surely not all.

So, what, apart from debris, did the Moon landing achieve? There is the geopolitical angle—the Space Race was a major battle in the long decades of the Cold War. The system that delivered the technical prowess and money required to win that battle dealt the other system a psychological blow. It is said the Cold War was won "without a shot being fired." Given the number of proxy wars it spawned around the world, that was never true, but another shot, the "Moonshot," played its role.

There are also the scientific achievements the wider Space Race underpinned: advances made by both sides. Computer science, telecommunications, microtechnology, and solar power technology were all rapidly boosted via the engineering required to get to the Moon and back. Modern portable water purification systems owe their lineage to those invented by NASA. So do the lighter breathing masks used by firefighters around the world, as well as their heat-resistant clothing. Laptops, wireless headsets, LED lights, and memory-foam mattresses? All can be traced back to the science of the Space Race, some directly.

But wireless headsets and breathing masks are mere minor details of history, and even the Cold War will eventually be consigned to an afterthought. It's estimated that about 110 billion humans have walked the surface of Earth. Almost all of them will have gazed at the Moon in wonder. But only twelve have walked there. Armstrong setting foot on what Aldrin called a scene of "magnificent desolation" is a moment for the ages.

PART II

Right Here, Right Now

Space Shuttle Atlantis takes off from the Kennedy Space Center, Florida, headed for the International Space Station on November 16, 2009.

3

THE ERA OF ASTROPOLITICS

The first day or so we all pointed to our countries.
The third or fourth day we were pointing to our continents.
By the fifth day we were aware of only one Earth.
—Prince Sultan Bin Salman al-Saud, astronaut

Many of us still think of space as "out there" and "in the future." But it's here and now—the border into the great beyond is well within our reach.

The Space Race was all about getting up and out. Now we're claiming what's there. And as more countries become spacefaring nations, history suggests there will be competition and cooperation along the way. That will inevitably mean "spheres of influence" and even claims on territory as the rivalries, alliances, and conflicts on Earth spill out into space. Both military and civilian players are already eyeing opportunities from the satellite belt all the way to the Moon and outward.

This is the era of astropolitics.

The great geopolitical theorists of the nineteenth and twentieth centuries, such as Admiral Alfred Thayer Mahan (sea power) and Halford Mackinder (land power), factored in place, distance, and supplies when assessing the limits of what a country could and could not achieve, and the impact of this on international relations. Valleys, rivers, and mountains create the conditions under which we trade and sometimes fight each other.

"Astropolitics" applies similar principles. Like geopolitics, its basis is in geography. Outer space is not featureless—it has regions of intense radiation to be navigated, oceans of distance to cross, superhighways where a planet's gravity can accelerate spaceships, strategic corridors in which to place military and commercial equipment, and land rich in natural resources. All this attracts the attention of the big powers, who will try to establish and maintain an advantage. And it raises important questions as countries prepare for the scramble for space. Which strategic locations in space are most useful? Which planets might have water or minerals? What is the density of their atmospheres? Is there a viable planet we could colonize?

An understanding of the geography of space is necessary if we are to understand astropolitics.

The geography of space begins on Earth, as first we need to find a way up. The costs and effort required have certainly lessened since the Apollo era, but if you want to be a spacefaring nation—or company—you need a serious amount of money and either rocket launch capability or access to a suitable part of the world that is willing to host you.

And so we start, literally, on terra firma, with the locations best suited for launching rockets. Think of these as the ports from which vessels set out on voyages. The most functional location for launch is one that takes maximum advantage of Earth's rotational speed for the quickest entry into space—thus using less fuel—which means somewhere close to the equator where Earth's rotation is fastest (about 1,037 mph). Thus, the US uses the Kennedy Space Center Launch Complex in Florida, as close to the equator as its borders allow, where the speed is about 914 mph. The EU has used French Guiana in South America, while Russia used Kazakhstan. Our planet rotates west to east, and so rockets are launched eastward to receive an extra boost from Earth's rotation speed, saving fuel and time. It's also important that the drop zone for rocket boosters is over mostly uninhabited areas—hence why many launch sites are positioned on eastern coastlines.

Ideally a country should also be large enough to have sufficient resources in expertise, engineering, technology, and rare-earth metals so that its space program needs no vital external support; and its population should be engaged in the project and believe strongly in the value of science and technological advancement. In addition, the bigger the country, the more of the sky it can see from its home territory and the easier it is to track satellites and spacecraft—friendly or otherwise.

Taking the above into account helps explain why currently China, the US, and Russia are the dominant powers, developing significant military and civilian presence in space. The EU would be able to join them if it took the long-term strategic choice to so do; India, too, has potential.

Having found a way off the surface of the planet, now we're heading up through the clouds and quickly zooming past the typical maximum cruising height for commercial planes—about 42,000 feet. Up another 197,000 feet and we are approaching space, defined by

NASA as beginning 50 miles above sea level—everything below that is Earth. However, the Swiss-based Fédération Aéronautique Internationale, which ratifies astronautical records, defines it as beginning at 62 miles. This is the Kármán line, the point at which a spacecraft will start to break free from Earth's gravity. We're entering cislunar space—covering the region between Earth's geosynchronous orbit and the Moon, 236,000 miles away. The term is from the Latin for "on this side of the Moon."

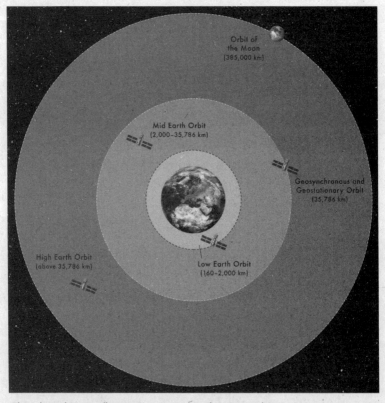

The orbits where satellites operate around Earth (not to scale).

When you reach low Earth orbit, from around 525,000 to 6.5 million feet above us, you might catch a glimpse of the International Space Station (ISS), which orbits at an average height of 1.3 million feet. This area has changed a lot since Sputnik went up, not least the politics. In 1993, a deal was agreed to among the American, Russian, European, Japanese, and Canadian space agencies to build a space station bridging political and cultural divides. In 1998, the Russians took the first piece up and two years later there was enough room for occupants to move in. Since then, more than 160 Americans and over fifty Russians have shared the living quarters and science labs with dozens of other astronauts, including eleven from Japan, nine from Canada, five from Italy, and four each from France and Germany. Other countries have sent people to contribute to the ongoing scientific work carried out there: Belgium, Brazil, Denmark, the UK, Israel, Kazakhstan, Malaysia, the Netherlands, South Africa, South Korea, Spain, Sweden, and the UAE. The record number of people present at any one time is thirteen. Mission controls in Moscow and Houston saw those men and women there and back, usually via a Russian Soyuz capsule. The ISS is a symbol of what can be achieved in space through cooperation. Sadly, it has almost reached the end of its life span and is due to be decommissioned in 2031; it will be crashed into a remote part of the Pacific known as Point Nemo, where it will sleep with the fishes.

But you might miss the ISS's descent, what with all the other traffic hurtling around. Low Earth orbit is an attractive piece of real estate because that's where most satellites operate. Without satellites, international communication networks and global positioning systems would not exist. Jam, spoof, or destroy these satellites and your grocery delivery van can't find you, emergency services are lost, ships drift off course, and a major industrialized economy such as the UK loses an estimated $1.2 billion a day. Their importance to modern life cannot be overstated and their function in the military is now key to modern warfare.

Modern satellites come in various shapes and weights, from small ones about the size of a Rubik's Cube weighing just 3 pounds all the way up to those weighing upwards of 2,200 pounds, the traditional workhorses of the industry. Most models have solar panels to derive power from the Sun as well as panels to protect the electronics from the intense heat. They all require a communication system, a computer to monitor a range of measurements including altitude and orientation, and a means of propulsion to course-correct if they are drifting out of the required orbit.

Satellites arrive in orbit after hitching a ride on a rocket that has been fired vertically to punch through the atmosphere as quickly as possible in order to reduce fuel consumption. Most then fly west to east, following the direction of Earth's rotation. Fewer satellites fly the north-to-south polar orbit because the direction of launch means more fuel is required. Those in polar orbit are mostly used for mapping, weather monitoring, and reconnaissance, and a complete orbit takes approximately one hundred minutes. The satellite observes the globe in segments, as they are both moving in a different direction, as if it were a giant pale blue satsuma. The entire surface can be mapped this way in twenty-four hours.

Satellites in the standard west-to-east orbit take between ninety minutes and two hours to orbit the planet, depending on how far away from Earth they are, spending only a few minutes over a target area on each pass. They tend to work in groups, or constellations, to create a "net" and often communicate with each other, as well as with the ground stations, to create permanent coverage. America's Global Positioning System (GPS) uses a minimum of twenty-four satellites distributed equally around the planet to achieve this.

Low Earth orbit is the region most commonly used for satellite imaging: being relatively close to Earth's surface allows clearer pictures. The detail that military-grade satellite cameras can capture, for example, is impressive. A civilian weather satellite might have a resolution of 3,300 feet, which means you can't see anything smaller than 3,300 feet in size—fine for measuring sea temperatures, not so good for identifying Jason Bourne walking out of a building. Anything above 164 feet is considered low resolution. Modern high-end military satellite resolution is thought to go down as far as 0.49 feet (or 6 inches), so now you can identify what brand of sunglasses Bourne is wearing. Commercial sale of this technology is not allowed on security grounds. If a satellite is being used for surveillance, then spotting it, or knowing when it is overhead, is very useful for those who prefer not to be watched. Some can be seen with the naked eye; others require inside knowledge to know their location.

Strategically, low Earth orbit is a potential "choke point." We're familiar with these on Earth; for example, the Suez Canal and Strait of Hormuz: places where sea lanes are narrow and can easily be blockaded. It's not an exact analogy, but it is a useful one. Just as you need to be able to defend your launch sites in order to venture into space, you also need to ensure you have access to the lines of communication provided by satellites in low Earth orbit, and also to be able to move through it on your way out to the "ocean" of the cosmos.

As we continue our journey upward, we need to avoid loitering in the Van Allen radiation belts—two doughnut-shaped areas extending out from Earth for thousands of miles containing high-energy particles trapped by Earth's magnetic field. Concentrations of radiation vary, but in places they are high enough to fry a spacecraft's electronics and, over time, break apart the chemical bonds in human cells.

At around 6.5 million feet up we enter medium Earth orbit, which goes up to about 117.5 million feet. Satellites here take twelve hours or so to go around the world. Many of them provide positioning and navigation services on Earth. These machines carry atomic clocks that measure time according to vibrations of atoms. They are said to be so accurate that, when linked with updates from atomic clocks on Earth, they wouldn't gain or lose a second over millions of years. The satellite sends a radio signal (at the speed of light) to a receiver on Earth, including one in your smartphone or car's sat-nav system. This works out your location as you move about so your car knows where it is and how to get somewhere else. Usually.

Onward and upward to high Earth orbit, starting with the region of geosynchronous and geostationary orbit, 117.5 million feet above Earth. The only difference between them is that a satellite in geo-synchronous orbit can circle the planet at any inclination, while a geostationary satellite always follows the equator.

Low Earth orbit is difficult territory for communication satellites because they move so quickly that it's hard for ground stations to keep track of them, but up here the speed of the satellite matches the speed of the rotation of Earth and so is above the same piece of territory all the time. If you could see one from Earth it would appear stationary. A single machine can see up to 42 percent of Earth's surface. Military communication and intercept satellites live here along with TV, radio, and some long-range weather satellites. It's busy, but much less so than low Earth orbit. Due to signal interference there are only so many "slots" there, and limited frequencies on which machines can communicate. The UN's International Telecommunication Union awards both the positions and frequencies so you can't just pull up and park there.

This is where the Americans hold their six dual-use Advanced Extremely High Frequency satellites that communicate with their war planes; with the British, Dutch, Australian, and Canadian militaries; and with the US nuclear early-warning system. The Russian early-warning Unified Satellite Communication System is in the same orbit, and it's thought that parts of China's BeiDou satellite system have similar capabilities.

Farther out into high Earth orbit is where many satellites go to die. As a satellite comes to the end of its natural life, onboard thrusters push it out of geosynchronous orbit, deeper into space, to ensure it is not a hazard to others.

It's getting busy above Terra and is destined to become more so. More than eighty countries have crossed the border and placed satellites in space, taken there by the eleven countries that have (or had) launch capabilities. The biggest players are the US, China, and Russia, with Japan, India, Germany, and the UK positioning themselves to be among the front-runners. Also claiming their place in the satellite belt are Tunisia, Ghana, Angola, Bolivia, Peru, Laos, Iraq, and dozens of other countries not usually associated with machines orbiting the planet. Many of these satellites are launched by private companies, not just states. According to the Union of Concerned Scientists there are currently well over eight thousand satellites hurtling around Earth, of which about 60 percent are active, and they are going to be joined by many, many more. There's plenty of room for hundreds of thousands of them, but with each new one the risks of collision and outright conflict increase.

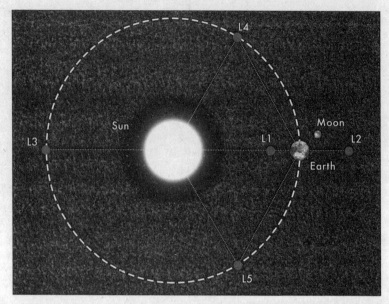

The Lagrange points of the Earth-Sun system (not to scale), advantageous positions to place satellites. These points exist in all two-body systems, including the Earth and Moon.

Farther out, other key areas for satellites are the Lagrange points. These are "parking lots" in space, places where the gravitational pull of two large masses that are orbiting each other is balanced equally between them. This means that a third, smaller body, such as a satellite or spacecraft, can "hover" at the sweet spot and move with them while using minimal fuel. Alternatively, in the future you could deliver a consignment of raw materials mined from an asteroid, or equipment needed to build a space station, to one of these points and be confident that it would still be there when you returned.

There are five Lagrange points in each two-body system, for example, the Sun and Jupiter, but the ones that concern us are those of Earth and the Sun, and Earth and the Moon. L1 in the Earth-Sun system may be 932,057 miles away, but it's home to SOHO—the Solar and Heliospheric Observatory, which keeps a continuous eye on the Sun from a safe(ish) distance. The James Webb Space Telescope arrived at L2 in 2022, and because the telescope faces away from the Sun, Earth, and Moon it has uninterrupted views of deep space. Minor adjustments, using almost no fuel, should keep it there for the next twenty years.

L4 and L5 are not in use yet, and few people care about L3 as it's hidden on the other side of the Sun. But it has been useful for sci-fi writers, who have imagined an exact mirror image of Earth there, an idea best captured by the 1969 film *Doppelgänger*, also known as *Journey to the Far Side of the Sun*. There are some nice touches in it: Earth's plucky astronaut thinks he's crash-landed back home until . . . he realizes the writing is backward, and worse—people are driving on the wrong side of the road! It's a bit like being in Russia.

Back in the real world (I think), in the Earth-Moon system, both L1 and L2 may become important as the locations for "gateway" space stations near the Moon. L2 in particular is on the far side of the Moon and as such provides "radio silence," meaning scientists can study the universe without interference from Earth communications. The strategic advantages of the L points suggest there could be competition for them. Fortunately, they are huge—about 500,000 miles wide—so there is room to spare, although space powers operating in these regions will keep a wary eye on one another.

L3 is less useful as it is on the other side of Earth to the Moon. L4 and L5 are also not currently being used, but because they are relatively close to Earth they have been discussed as potential places to park future space colonies. In the 1970s and '80s there was a group called the L5 Society, which sounds like it might be a bit weird—as well as smacking of favoritism—but was in fact formed by serious scientists to promote the ideas of a physics professor at Princeton University, Gerard K. O'Neill. They also had a sense of humor, as shown by an early missive: "Our clearly stated long-range goal will be to disband the Society in a mass meeting in L5." In 1987, the society and its 10,000 members merged with the larger National Space Institute and is now called the National Space Society.

The final stop on our cislunar tour is the Moon itself—238,855 miles distant, a mere 1.3 light-seconds away—the time it takes light to travel from the Moon to our rock. Driving at 62 mph, it would take less than an hour to get from Earth into space, but then another six months to get there. The quickest journey to date was that of the New Horizons spacecraft—eight hours and thirty-five minutes—but most crewed journeys take about three days.

The Moon's surface and shape have now been mapped. It is a stunning place, with mountains, ridges, valleys, plains, and huge caves. Its surface area is just under 14.6 million square miles, slightly larger than Africa. For the better part of a billion years the Moon was bombarded by meteorites, some so large they created the multi-ringed basins and mountains we can see today with the naked eye. We can also see light and dark areas—the highlands and the "marias," from the Latin for seas, which is what early astronomers thought they might be. In fact, the meteorite impacts caused volcanic activity, resulting in lava flows on the surface. They are darker because the high iron content of the volcanic rock reflects less sunlight than other areas. By the time Apollo 11 landed in the Sea of Tranquility in 1969 we'd worked out it wouldn't be a splash landing. If you look at a full moon on a clear night (from the Northern Hemisphere), you can see the 541-mile-wide Sea of Tranquility just to the east of its center. The rest of the surface is called terrae (land) and contains the mountainous regions, some of which are 3 miles above average elevation.

Recently evidence has been found for deposits of metal oxides in some of the large craters. It's thought the meteorites may have excavated the material from beneath the surface. If so, it's probable there will be large concentrations of metal oxides deeper underground. And it's believed the Moon contains reserves of silicon, titanium, rare-earth metals, and aluminum. Humanity is destined to spend more time there, digging in pursuit of these metals, which are used in vital modern technologies. Many countries have the incentive to go after them, especially those that don't want to rely on China, which currently holds a third of the world's known reserves.

There is also the potential for a serious amount of energy, enough to power human communities on the Moon and be exported back home. The potential lies in helium. The name of this rare and noble gas is derived from the Greek word "helios," meaning "sun"—because that's where it was first detected. The isotope helium-4 makes up more than 99 percent of the natural helium found on Earth. Jolly useful stuff it is too. It inflates balloons at children's parties, for example, not to mention car airbags, and it plays a role in cooling parts of magnetic-resonance imaging systems. But it's not helium-3, and that's what we're after.

Theoretically, helium-3 can be used to create nuclear fusion—the holy grail of energy production, as it would produce higher amounts of energy than nuclear fission but is non-radioactive. On Earth only about 0.0001 percent of helium is helium-3, but on the Moon there may be a million tons of the stuff. This is because our satellite lacks an atmosphere and so for billions of years solar winds carrying helium-3 have saturated the surface.

Ouyang Ziyuan, China's distinguished chief scientist in its Lunar Exploration Program, believes that if helium-3's power can be harnessed it will "solve humanity's energy demand for around 10,000 years." That's forward-thinking, but it's also thinking about the present energy crisis and climate change. Scientists can't give exact figures about how much helium-3 is required to make X amount of energy, but estimates suggest one ton might be the equivalent of 50 million barrels of crude oil.

Scientists have been working on nuclear fusion reactors for forty years and there are basic prototypes in existence, but barring an unexpected breakthrough the technology required to achieve it probably belongs to the next decade, not this one. So may the technology required to mine on the Moon, but the process has begun.

It is also believed there are deposits of water. Some 1,678 miles south of the Moon's equator lies the South Pole–Aitken basin, which is 1,553 miles wide and 8 miles deep. Within it are towering mountains, some of which are bathed in sunlight for up to 80 percent of the time due to the tilt of the Moon's axis. In the late 1800s, it was theorized that these mountains might be permanently lit and were dubbed "the Peaks of Eternal Light," but it now seems that even the highest at times fall dark. There are craters near them, however, that are so deep that sunlight, shining at a shallow angle, never reaches their lower points. These permanently shadowed locations are the coldest places observed in the solar system. Temperatures as low as -396°F have been recorded, which is colder than the surface of Pluto. In the freezing caverns are ice crystals, and in the ice are oxygen and hydrogen; from that you can make rocket fuel.

If you can get the ice out of the ground, you pass electricity through it and it divides into liquid oxygen and liquid hydrogen. Sure, there's more to it, but you get the idea, and given that some estimates suggest there are 1.1 billion tons of water ice at each lunar pole, it may be a very good idea. Launching a rocket from the Moon requires a fraction of the fuel needed to escape Earth's gravity, and so once the infrastructure is in place, an Earth–Moon trip would not need to carry enough fuel for the return journey if there are supplies on the "orbiting garage." NASA's giant Space Launch System (SLS) rocket is designed to burn 802,500 gallons of fuel to get from Earth to low Earth orbit, which is the equivalent of draining 1.2 Olympic swimming pools in about nine minutes. This is one of the reasons why the Moon will also be useful for launching long-distance missions from bases on its surface.

And the geography past the Moon? The limit is infinity, which is to say there is no limit. But for the foreseeable future crewed spacecraft won't need a map that extends farther than Mars, and even that probably won't be until the 2030s at the earliest. Compared to the vast distances in space, the planets in our system settled relatively close to each other, but while we can get machines to all of them now, they have remained beyond our ability to visit. Jupiter averages 365 million miles from us, Saturn 996 million miles, and Neptune 2.7 billion miles. However, the phrase "it might as well be on Mars" is becoming out of date. The first spacecraft to do a flyby of Mars was NASA's Mariner 4, which reached the planet in 1965. After that other craft orbited Mars until 1971, when the USSR landed Mars 3, which transmitted a fuzzy signal for fourteen brief seconds, then cut out, never to be heard from again. Five years later NASA's Viking 1 arrived, landing on the western slopes of the "Golden Plain," and began sending back the first photographs of the surface. Now Mars is one of the best-mapped planets in the solar system and the only one rovers have explored.

The latest spacecraft can get to Mars in about seven months, and a crewed trip is on the horizon. The entrepreneur billionaire Elon Musk, CEO of SpaceX (full name Space Exploration Technologies Corp.), says he intends to send humans to the surface of Mars in this decade and that the journey time will be eighty days or less. Technological change moves faster than ever, but those time frames still seem north of ambitious. Timing will be crucial. The average distance to Mars is 140 million miles, but, as with all the planets, distances change according to the cycle of orbits. The closest possible distance is about 34 million miles, the farthest 250 million. The mission will probably be launched at a time when the Red Planet is closer to us than average. This means that Mars is in our sights so far as boots on the ground are concerned. And from there the plan is that craft would be able to refuel and "planet hop" to other targets, eventually reaching the outer limits of our solar system. But for now, that's still over to the robots for the next few decades at least.

The Moon, however, is within our reach, and the main space-faring nations are keen to set up shop sooner rather than later. Yes, mining and processing on the Moon will be incredibly difficult; yes, nuclear fusion from helium-3 may only be theoretical; and yes, time frames and budgets will slip, but can you stand by and watch your rival get so far ahead of you that if theory becomes fact, you are out of the game? Helium and water are not practicable renewable resources; you can't wait another billion years for waves of solar winds to replace what is dug out and heated—it's first come, first served. The financial model does not yet make sense, but we didn't go to the Moon the first time for financial gain. The exploration, and exploitation, of the New World shaped the last five hundred years of history. What lies above and beyond has this potential.

The challenges will be taken on for various reasons—prestige, commercial, and strategic. A successful colonization of the Moon will give a country, or an alliance, advantages similar to those enjoyed by maritime powers in previous ages. A dominant power will be able to stymie the ambitions of others by occupying the territory and attempting to police it. Its satellites will enjoy a direct line of sight down to geostationary and low Earth orbit. Those who pave the way will set parameters others may be expected to follow. The first to establish themselves will be the first to access the potential wealth of the Moon and the ability to ship some of that wealth back home.

If a space superpower could dominate the exit points from Earth and routes out of the atmosphere, it could prevent other nations from engaging in space travel. If it dominates the Moon it can keep its riches and be the only power using it to travel farther. And if it dominates low Earth orbit, it could command the satellite belt and use it to control the world.

One of the world's leading theorists of astropolitics is Everett Dolman, professor of military strategy at the US Air Force Air Command and Staff College and author of the prescient *Astropolitik: Classical Geopolitics in the Space Age*. Professor Dolman coined one of the field's best-known maxims: "Who controls low Earth orbit controls near-Earth space. Who controls near-Earth space dominates Terra. Who dominates Terra determines the destiny of humankind."

Because of this, the temptation to dominate regions of space is growing. The three main powers are now in competition, locked in an arms race to ensure that neither of the others can call the shots. And that is causing other countries to consider what their own military options might be. Nations such as Japan, France, and the UK have all announced their own military space commands.

There is a familiar cold logic to this. If you have longer-range arrows (see the Battle of Agincourt for details), I'll develop better shields while I work on my range. In previous times, commanders would not send soldiers into battle without the means to defend themselves or attack the enemy—and in this age, satellites are a crucial part of warfare and a vital part of the early-warning systems countries use to detect the launch of nuclear weapons. It follows that losing such a satellite will render a country vulnerable; being denied access to the orbit belts would make life very challenging indeed. No country relying on its satellites for going to war, or for early warning of being attacked, will choose to leave them defenseless and forgo the ability to hit an enemy satellite.

The "laws" we currently have for activity in space are little better than guidelines. Technology and changing geopolitical realities have overtaken them. With an increasing number of space-based platforms for military and civilian uses—mining, solar energy projects, scientific work, and space tourism—space is becoming a congested twenty-first-century environment requiring twenty-first-century laws and agreements.

The idea that space is a global commons is disappearing. The stakes are high. We need a new set of rules, and a better understanding of the space they govern. There are 8 billion reasons why. Every human on Earth has a stake in a rules-based space order, and in global cooperation on cosmic issues. Without this we may end up fighting over the geography of space, just as we have done over the geography of Earth.

4

OUTLAWS

From out there on the moon, international politics look so petty.
You want to grab a politician by the scruff of the neck and drag him a quarter
of a million miles out and say, "Look at that, you son of a bitch."
—Edgar Mitchell, Apollo 14 astronaut

B etween the third rock and the Sun is a hard place. It has a challenging geography and is a hostile environment, but it also contains riches untold. Like so many regions humans have encountered before with similar characteristics, it is virtually lawless. It is space, and it needs space laws.

But that's no easy feat. Laws and agreements are difficult enough on Earth, where there are clearer boundaries and borders, and established precedents. What's more, in space it's not in the interests of the big powers to give up their advantage.

Existing space laws are horribly out of date and too vague for current conditions. Most were primarily products of the Cold War, negotiated by the main players. They are no longer fit for purpose. The Outer Space Treaty (1967), for example, upon which most of the rules governing the use of space are based, says: "Outer space, including the moon and other celestial bodies, is not subject to national appropria-

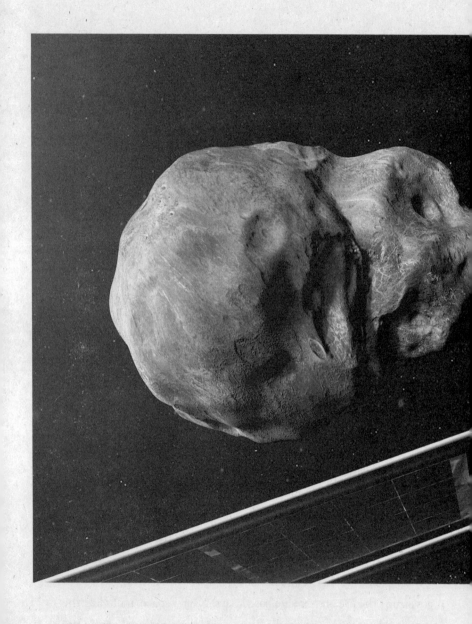

Illustration depicting NASA's Double Asteroid Redirection Test (DART), targeting the asteroid Dimorphos, September 2022.

tion by claim of sovereignty, by means of use or occupation, or by any other means," and exploration "shall be carried out for the benefit and in the interests of all countries, irrespective of their degree of economic or scientific development, and shall be the province of all mankind." If a country were to build a base on the Moon with defined areas where it is not safe for other nations to operate, would that constitute occupation and/or sovereignty? If one country mines the Moon for resources to be sold on Earth—is that in the interest of all mankind? The treaty also prohibits placing weapons of mass destruction in space but makes no mention of conventional weapons. And in any case, it is a document without enforcement measures. The Moon Agreement (1979) is similarly outdated and has too few signatories to be effective—it's worth noting that it hasn't been ratified by the US, China, or Russia.

Such treaties fail to cover the changes in technology available to nation-states, and don't reflect the fact that dozens of low- and middle-income countries are now in the game but had little input when the rules were first drawn up. As Professor John Bew, foreign policy adviser to 10 Downing Street, says: "Space is one of the new frontiers of the international order where the balance of power is contested and the rules have not yet been fully written."

In place of these relics of another age, a series of nonbinding ad hoc agreements have emerged. The Artemis Accords (2020) is the best example. It purports to set out updated guidelines for activity on the Moon. Some parts are in harmony with the Moon Agreement: both promote the rule of law in exploration, agree on providing assistance for all astronauts and spacecraft regardless of nationality, and call for the release of scientific data gathered on the Moon.

However, there is a fundamental difference between the two

in that the agreement promotes a multilateral, indeed global legal framework for the Moon, whereas the accords are a series of bilateral agreements, and the text is mostly authored by the US and reflects its approach to space laws. Some of the "updates" clash with the principles and philosophies laid out in the core provisions of the Moon Agreement—the Americans do not embrace the idea that lunar activities should be the common heritage of, and of benefit to, all humanity, for example.

Hence by joining the accords, the member states have effectively accepted the US legal approach to lunar law and—in the wider context—space laws. The original signatories were Australia, Canada, Japan, Luxembourg, Italy, the UK, the UAE, and the US; since then Romania, Ukraine, South Korea, New Zealand, Brazil, Poland, Mexico, Israel, Saudi Arabia, France, Singapore, and others have joined. But more than 160 other countries have not, and China and Russia were specifically excluded. Congress has banned NASA from cooperating with China, and Russia was frozen out after it was accused of tracking US spy satellites in a dangerous manner.

In Greek mythology Artemis is the goddess of the Moon, twin sister of Apollo. The Artemis countries don't have lofty aspirations to be goddess of the Moon, but they certainly are ambitious. The mission is to land humans on the Moon within a few years and then begin to build permanent structures there by the end of the decade, with a view to achieving habitation early in the 2030s.

Signatories to the Artemis Accords buy into the idea that they clarify the legal basis for establishing a presence on the Moon and mining it for rare-earth materials, water, and hydrogen. It states that extracting resources does not inherently constitute national appropriation—in other words, the country doing the mining does not own the territory it's working on. In practice, though, this means it will be "first come, first benefits." China is likely to arrive on the Moon relatively soon after the signatories to the accords. If it is discovered that there are limited areas where mining is feasible, they will be bumping up against the competition—which by that time will already have staked out their claims. Less-developed nations will also miss out on what the Outer Space Treaty says is "the province of all mankind."

Section 11 of the accords has the noble aim of "Deconfliction of Space Activities." To achieve this, anyone who is setting up shop on the Moon will provide "notification of their activities." Those activities will take place in a "safety zone," defined as the area in which the activities of another nation "could reasonably cause harmful interference."

It gets worse, or perhaps better, depending on whether you are a space lawyer on an hourly rate. Apparently, the safety zones will change over time and so the "operating Signatory should alter the size and scope of the corresponding safety zone as appropriate." But fear not—signatories will let the public know all relevant information available "as soon as practicable and feasible." Phew! That's a relief. But what's this? Oh, they'll only do that "while taking into account appropriate protections for proprietary and export-controlled information." You could drive the Starship *Enterprise* through the gaps in this legalese, especially as most countries in the world have not signed on to it. Even if they had—define "reasonably," "harmful," and "interference."

So, let's rewrite that section: "After solemnly acknowledging, affirming, and committing to the principle of free access to all areas of celestial bodies, the Signatories affirm their right to mark out boundaries where others cannot go in case they get in the way. The Signatories will define the boundaries and reserve the right to change them. The Signatories commit to transparency in these matters except when they choose not to."

There. Fixed it. It's not that the original text is right or wrong, but it's got more holes in it than the surface of the Moon.

Supporters of the accords argue that because it is agreed that the Moon shall be used only for peaceful purposes, "safety zones" are not a problem. However, what constitutes "peaceful" is not defined either, and what if my definition differs from yours? In 1959, in relation to the Antarctic Treaty, Russia defined "peaceful" as "nonmilitary." However, the US interprets it as meaning "nonaggressive," which allows for military activity if it is not aggressive. These two interpretations, on aggression principle, will keep those space lawyers employed for years to come. Clauses in the Outer Space Treaty already allow for military personnel to work in space for scientific or peaceful purposes. Once you've built "facts on the Moon," however, if another non-Artemis country entered your "safety zone," it would be easy to argue that you needed defensive weapons, not for aggressive purposes, of course, but to ensure peace. And once you've got defensive weapons, I want them as well. Just for defensive purposes, of course . . .

It's also not too giant a space hop to get from "safety zone" to "spheres of influence," another term with a hazy legal definition, but one that essentially refers to a region where a nation claims some form of exclusivity, whether economic, cultural, or military. Our earthly obsession with such spheres has contributed to conflict down the ages, so exporting them to space might not be the best idea.

There's more. With reference to private companies that might be working with a nation-state, each Artemis country "commits to taking appropriate steps to ensure that entities acting on its behalf comply with the principles of these Accords." However, a major US company operating on the Moon could point to the 2015 Commercial Space Launch Competitiveness Act, which allows American citizens to privately "possess, own, transport, use, and sell the resources" obtained in outer space. Given that the laws of a country do not apply outside its borders, other nations would have grounds to protest, but this too could become complicated.

Section 9 of the accords introduces a new concept: that of preserving outer space heritage, but doesn't define what that might be or how to guarantee it. This throws up the scenario of the US unilaterally declaring the site of the Apollo 11 landing, Neil Armstrong's footprint, and the American flag as being of historical value and classifying the whole area as an American safety zone. Armstrong's footprint is indeed of historical value, but it is another matter to set up what might become de facto "law" on a unilateral basis.

It's worth noting that more countries have now signed the accords than the original Moon Agreement, which is often cited when discussing legality. If enough countries consider an agreement to have the same weight as international law, then as the years pass, and the practices laid out in the text become ingrained, countries begin to treat them as if they are law. If a clear majority of countries have ratified a document, it's generally agreed to be the international global legal standard. An example is the United Nations Convention on the Law of the Sea (UNCLOS), which has established a legal framework for activities at sea and what constitutes maritime borders. Originally put forward in 1982, it finally came into effect in 1994 when it reached 60 signatories, and now has 157. Several major countries, notably the US and Turkey, have not signed it, but that does not stop it from being regarded as the global "constitution of the oceans."

Just as UNCLOS is now cited in maritime disputes, it's reasonable to posit that in the 2030s the Artemis countries, whose numbers will grow, will cite the accords in a dispute with Russia or China about regions on the Moon. But just as Turkey does not accept UNCLOS definitions in disputes it has with Greece about oil and gas reserves in the Mediterranean, so we can hardly expect Beijing and Moscow to adhere to Artemis definitions.

In 2020, the then head of the Federal Space Agency (RSA), Dmitry Rogozin, described the accords as akin to an "invasion" of the Moon that could turn it into "another Afghanistan or Iraq." The following year Russia and China signed their own memorandum of understanding to build a Moon outpost, to be called the International Lunar Research Station, and said the project was open to other countries to join.

A new set of treaties is therefore required to deal with the new realities created by technology, and not just to prevent "safety zones" from turning into war zones. The problem is that they need the buy-in of all players. And given that as of yet we can't even agree whether space starts—and a nation's sovereign territory ends—at 50 or 62 miles up, there's a long way to go.

It's complicated enough navigating all these issues between Earth's nations, but there are plenty of other things our outdated space laws need to address. What even counts as activity in space, for example? If a country uses a space-based satellite to control a drone on Earth, which then fires a missile at a military target, does that breach the Outer Space Treaty?

If the satellite used was a commercial one, does that mean the entire satellite system of that company can now be treated as a weapon? In 2004, during the Iraq War, 68 percent of US munitions fired were guided by satellites, of which 80 percent were commercial. If Iraq had possessed the capability, would it have been within its rights to fire at those satellites? In 2022, a new dynamic entered the argument.

During the early days of Russia's invasion of Ukraine, the city of Irpin lost connection to the internet after all twenty-four of its base stations went offline after most were hit by missiles. Two days later connection was restored. Elon Musk's SpaceX had sent Starlink high-speed terminals to the city to connect with the advanced Starlink satellites in low Earth orbit. More than ten thousand of what Starlink engineers call Dishy McFlatfaces were then distributed around the country. Most were used by ordinary people, but the Ukrainian military maintained networked contact, allowing it to retain command and control of facilities, including for their drones, which sent targeting information to commanders.

The Russians tried to jam the signal between the terminals and the satellites, but SpaceX quickly worked out how to evade this. This was all noted in Moscow and Washington. Dave Tremper, director of electronic warfare at the Pentagon, said, "We need to be able to have that agility," while Dmitry Rogozin at the RSA complained that Starlink was functioning as an arm of the Pentagon. If that was true, could Russia have legitimately attacked Starlink satellites? After all, it was being used as part of the process to kill Russian soldiers. You could argue that SpaceX was a third party from a country that was fighting a proxy war. Another realistic current scenario: What would China do if the Communist Party was facing a potentially successful uprising against it and Starlink beamed down internet links bypassing the Great Firewall, thus allowing citizens to organize at a national level?

Plans are being put in place to deal with situations such as this. In 2019, NATO added space to land, air, sea, and cyberspace as an operational domain, and the following year agreed to establish a space center, which opened in Ramstein, Germany, in 2021. Personnel are drawn from various NATO members, with a brief to coordinate the data from member states that have space commands on navigation, weather, and potential threats to any NATO country. Despite input from the French and British, the alliance still relies heavily on the US for reconnaissance and spotting targets, just as it does on the ground for a lot of conventional war-fighting capabilities.

NATO's 2021 summit included a little-noticed statement expanding the Article 5 mutual defense clause to include space. The statement was carefully worded: "attacks to, from, or within space . . . could be as harmful to modern societies as a conventional attack. Such attacks could lead to the invocation of Article 5." A decision would be taken on "a case-by-case basis."

The cautious language—"could" and "case-by-case"—reflects that we've entered new territory. This is not a minor detail. It's easy to classify firing missiles at a NATO country as an act of war, but what about firing a laser beam that fries a commercial satellite? The act would not take place in sovereign territory and there would be no human casualties. Is it worth declaring war? Is Spain, for example, going to take up arms over one of Elon Musk's satellites being hit as it passes above Kenya? Probably not, and even in this scenario there is complexity. Article 6 defines the operational territories of the thirty NATO states and talks about attacks on them "when in or over these territories." This suggests that an attack on an object when it is orbiting an uninhabited part of the Pacific is not necessarily a trigger for Article 5, but still leaves it unclear whether hundreds of miles up in space counts as "over" sovereign territory.

Hence the "case-by-case" stance. It gives NATO a position of strategic ambiguity about what it might do, rather than obliging it to trigger a military response. However, whatever the definitions, it's unlikely that geographical restrictions would apply in the event that one of the US's early-warning satellites was knocked out.

The alarm caused when a Chinese surveillance balloon flew over the United States in early 2023 also throws up questions about potential scenarios to which there are not yet answers. American satellites tracked the balloon from the moment it took off from Hainan Island and headed toward Guam. Senior officials at Northern Command were informed it was approaching Alaska on January 27 and followed it closely for the next few days. On February 4, a Sidewinder missile fired from an F-22 jet shot it down off the coast of South Carolina. The pilot radioed in, "Splash one! . . . The balloon is completely destroyed."

During its overflight of Montana, the balloon was above the Malmstrom Air Force Base, which is home to nuclear missile silos. Its equipment included high-resolution cameras and technology allowing it to capture electronic data and voice communications, including those on cell phones. There is some thought that it also had the ability to transmit captured information up to China's spy satellites, which could send it on to Beijing as part of an overall picture about America's radar and weapons systems function.

Shooting down the balloon was not an act of war. It had clearly violated US sovereign airspace even though China claimed "force majeure"—that an act of nature had pushed it into US territory.

But suppose US intelligence learned in real time that it had gathered above top-secret information fundamental to the security of the country, and was sending it to a satellite. It's possible the order would have been given to at least blind the satellite or fry its electrics in an attempt to contain the security breach, or in extremis, even shoot it down. These are the types of scenarios security officials now have to think about and give their superiors options for.

The presence of corporate and private enterprise in space also raises all sorts of questions unrelated to military activity. Which of Earth's laws would apply to their ventures—and how would they be enforced? Let's imagine that Space Mogul Frankenstein builds an artificial human from living tissue while on board Space Station Shelley. International treaties between countries on Earth may have banned creating such a human, but Space Mogul Frankenstein is not a country, and Space Station Shelley is not on Earth, so who is going to stop him and how?

Outlandish, yes, but plausible. Scientists on board the ISS have already created living tissue using a 3D printer and bio-ink while working in the BioFabrication Facility. Similar work happens on Earth, but here the amount of tissue that can be built is limited because gravity collapses the delicate material. In space, scientists can print a sort of tissue scaffold and then add to it. They are on their way to being able to print human organs. Given the scarcity of organ donors on Earth (not to mention Mars), these scientific breakthroughs might be to humanity's advantage; however, the legal framework governing such projects in space is vague.

Aboard the ISS it is generally understood that the national law of the country from where a scientist comes holds sway. For example, an invention made in the Japanese Experiment Module (JEM) is accepted as having occurred in Japan. But that is due to an agreement signed by participating countries.

On the more macabre side of things . . . In the unlikely scenario of a Japanese astronaut killing a Japanese colleague in the Japanese module, the law is clear. The Outer Space Treaty states that legal jurisdiction is retained by the country that has registered the object launched into space. This is similar to the laws about the registry of ships and aircraft. But it would get complicated if murder most foul involved two people from different countries and took place in a connecting corridor—and murkier still if it happened outside the ISS during a spacewalk.

What about murder on the Orbital Express en route to SpaceTel—a million-star, two-hundred-room hotel orbiting the Moon? Or even in the hotel? It's much less straightforward if SpaceTel is owned by a private Indian company whose head office is in the Seychelles, with its parts made in Japan, carried up by rockets launched from Kazakhstan, the US, and China. Good luck with that, Space Inspector Poirot.

There are currently no easy answers to any of this, but it's worth noting that Canada has already moved to change its laws to allow its Criminal Code to extend to the surface of the Moon.

The only legal cases to date are both more prosaic and easier to unravel than the above scenario. In 2019, NASA astronaut Anne McClain was accused of accessing her ex-wife's bank account while living on the ISS. NASA investigated and found the accusations to be baseless; McClain's ex-wife was later charged with making false statements to federal authorities. Less seriously, Apollo 13 astronaut

Jack Swigert forgot to file his tax returns and remembered the error when he was in space. "Houston," he said, "I have a problem." Houston laughed and Swigert was granted an extension by the Internal Revenue Service on the grounds that he was "out of the country."

What if humans settle on a whole new planet; whose laws should hold sway there? Would it be governed from Earth? It's probable that colonies will eventually want to throw off the shackles of the "Mother Planet" and develop their own systems of self-government. The farther away you go, the harder it will be to enforce earthly laws. As we've seen, SpaceX intends to take humans to Mars. One of many things SpaceX does is provide broadband services via Starlink. Starlink has its terms of service, which includes the following paragraph: "For services provided on Mars, or in transit to Mars via Starship or other colonization spacecraft, the parties recognize Mars as a free planet and that no Earth-based government has authority or sovereignty over Martian activities. Accordingly, Disputes will be settled through self-governing principles, established in good faith, at the time of Martian settlement."

Self-governing principles? Who's in charge of this government and its principles? I smell a Muskrat.

The British academic and space expert Dr. Bleddyn Bowen pulls no punches in his response to the small print: "As I understand it Starlink has no legal right to put that in their terms of use because UN authority exists on Mars. The second part, about self-governing principles and good faith, is extremely politically naive and shows a typical ignorance of politics from technical/scientific communities that I see too often, sadly."

As seen above, Article II of the Outer Space Treaty says: "Outer space, including the moon and other celestial bodies, is not subject to national appropriation." Article III states that countries can only operate in space "in accordance with international law." In response to the point that Mr. Musk is not a country and therefore not bound by these rules, you could argue that the treaty also says that countries "bear international responsibility for national activities in outer space"—and you'd be right. But by the time SpaceX has launched from Honduras, aimed rocket-boosted lawyers at you, and switched headquarters from the US to Panama, Mr. Musk will already be the sheriff of County Mars. How the ultra-rich intend to govern their "colonies" remains to be seen.

As Dr. Bowen puts it, "Will billionaires run their 'colonies' the way they run their factory floors, and treat citizens like they treat their lowest-paid employees?" He's also uneasy about the term "colonies": "Is a word associated with genocide, corporate exploitation, ecological catastrophe, slavery and racism a word we want to use for a 'better future' in the cosmos?"

Once adequate, if sometimes vague, multilateral treaties, codes of conduct, and confidence-building measures have not evolved to keep pace with the emergence of private companies such as SpaceX, Virgin Galactic, and Blue Origin, and the lesser-known i-Space in China and Arsenal in Russia.

If a private company or individual is indeed able to establish a settlement on another planet, the "rulers" of these new off-Earth "colonies" will need oversight of the limits of their authority. There are several sci-fi novels with this story line at the heart of their plots, but in the real world, if we want to avoid a cosmic version of the East India Company, with its private army and de facto occupation of parts of India, we need laws fit for purpose.

There are other, more immediately pressing issues that also require international collaboration. A big one is space debris. Everett Dolman agrees that a new set of treaties is required to deal with this as a priority: "Debris is the number-one problem today. All spacefaring states have publicly advocated for mitigating and even reducing debris. The problem is that the proposals always clearly favor one side's interests."

NASA estimates there are more than 27,000 pieces of debris in orbit around Earth that are larger than 4 inches in diameter (roughly the size of a grapefruit). There are another 500,000 sized between .4 and 4 inches (a tennis ball is about 3 inches) and, in total, about 100 million bigger than .04 inches. Most pieces of debris may be small, but they are traveling at 17,504 mph, which would be troubling if you came in contact with one. A .4-inch fragment traveling at that speed can create as much energy as a small car crashing into you, or your spaceship, at 25 mph.

The sheer volume of satellites orbiting Earth means this problem is only going to get worse. SpaceX aims to launch 40,000 satellites for its Starlink broadband service; a new start-up called Astra has filed an application for 13,600; and Amazon wants 3,236. And that's just the Americans. Experts believe there could be a minimum of 50,000 by 2050—but there may be as many as 250,000 satellites in orbit by then.

More satellites will inevitably mean more debris. The more debris, the more risk of the Kessler Syndrome. The scenario is that the amount of junk in orbit reaches a point where collisions become frequent. This leads to a catastrophic cascade, resulting in a cloud of debris smashing into the Hubble Space Telescope, which then takes out a passing space shuttle before heading for the ISS. You might remember this being part of the story in the 2013 sci-fi film *Gravity*—

but it got the plotline from former NASA scientist Donald Kessler, who laid out the idea in a 1978 paper. In Kessler's version, the cascade continues until all satellites have been destroyed and the ring of debris in low Earth orbit makes it impossible for spaceships to leave our planet.

The Kessler Syndrome is a projection, but the current threat posed by debris is not hypothetical. On numerous occasions the ISS has had to fire its thrusters to avoid being hit by debris to maintain orbital altitude. Spacecraft have also hit each other in orbit. The best-known incident was in 2009 when Russia's inactive Cosmos 2251 communications satellite hit Iridium 33, an active US-based Iridium Satellite 497 miles above Siberia. Some 2,000 pieces of debris measuring at least 4 inches were added to the detritus circling Earth.

Efforts to reach an agreement for the reduction of debris are under way, but there are numerous complicating factors. A major one is that the creation of debris is not just accidental. A satellite is a tempting target for a number of reasons, and there are various anti-satellite (ASAT) weapons designed to target moving objects at high speed many thousands of miles above Earth's surface. The US first tested anti-satellite weapons in 1959. The program was continued by President Kennedy and subsequent presidents, culminating with Ronald Reagan's Strategic Defense Initiative, known as Star Wars. Naturally the Soviets were working on similar programs. They even installed a "self-defense" rapid-fire gun on board one of their Salyut space stations and in 1975 test-fired bullets into the atmosphere. It may not have been quite the "death ray" of sci-fi films, but it was certainly a first in space. The gun had its limitations. To aim it, the entire twenty-ton station had to turn toward a target, then fire its thrusters at the same time as the weapon to prevent the recoil sending the craft off into the unknown. It was also wise not to fire crosswise to its orbit as that would have resulted in the spacecraft shooting itself in the back. The only test known to have happened was done remotely after the cosmonauts had left.

Things have moved on a lot since then. There is now a whole array of precision weapons that can take down satellites—from either Earth or space. These include ballistic missiles, lasers fired from Earth to geostationary orbit, high-powered microwaves, and cyberattacks. There is even the potential to spray chemicals at a satellite's cameras to "blind" them, while the hydraulic arms on "space-cleaning" satellites, which are designed to grab debris, can easily be transformed into hostile weapons used to throw another satellite out of orbit.

In 2007, China used a ground-launched ASAT to destroy one of its own non-operational weather satellites 536 miles above Earth's surface in what looked like a test to see if they could do it against an enemy satellite or even spacecraft. A ballistic missile carrying a "kinetic kill vehicle" (KKV) was fired from the Xichang Satellite Launch Center in Sichuan Province. KKVs are sometimes called "smart rocks" because they don't have warheads that explode; instead, they simply smash into targets in a "hit-to-kill" attack.

The science of the destruction is the relatively easy part. The impact of the attacking vehicle needs to create a level of kinetic energy higher than the cohesive energy of the target, and therefore blow it apart. The difficult bit is arranging the collision at the speed required and, of course, hitting the target. The attack vehicle is not in orbit. It travels through space on a ballistic arc at a few miles per second, while its control systems track the speed and direction of the target, which is traveling even faster. Even the smallest deviation in the vehicle's trajectory, or a minute miscalculation of the target's speed and direction, and the attack will miss the target. If it hits, the effect is devastating.

The KKV in the 2007 test is thought to have weighed about 1,323 pounds, and collided with the satellite at a combined relative velocity of 19,884 mph. At such speeds solid objects behave like liquids and the two machines effectively passed through each other, creating a dust cloud that contained thousands of tiny pieces of metal. Other spacefaring nations were less impressed that the resulting debris, consisting of over 35,000 pieces larger than .4 inches across, then went racing around low Earth orbit, and many are still there. More debris was created in this test than by all previous incidents in the history of space travel.

Lessons have not been learned. In 2021, Russia used a direct-ascent hit-to-kill ASAT test to destroy one of its own satellites. Other countries have done the same, but the manner in which Moscow carried out the mission was beyond reckless. The satellite was blown into more than 1,500 pieces of metal, which immediately began to hurtle around the world—in the same orbit as the ISS. The seven people on board, four Americans, two Russians, and a German, were ordered to move into their docked spaceship capsules for two hours to allow a quick getaway, which in the end turned out not to be necessary.

U.S. Space Command issued a statement: "Russia has demonstrated a deliberate disregard for the security, safety, stability and long-term sustainability of the space domain for all nations." Numerous countries including Japan, South Korea, and Australia agreed. Russia didn't. Defense Minister Sergei Shoigu said it was a routine procedure allowing better Russian deterrence against American aggression and that there had been no danger to the ISS.

ASATs aren't the only way to take down satellites. All players will continue to develop their electronic warfare capability—work is being done on hacking into satellites' systems and taking control of them, denying the owners access or simply jamming them. However, it's unlikely they will rely solely on electronic warfare due to fears that rivals will still be developing kinetic weapons, leading to ever more debris. To combat that, a comprehensive treaty banning ASATs is required. But even if it does look "doable," it is difficult. The detail is devilish. You can't simply write, "We agree to ban ASATs." As well as the ground-based element, the text needs to define and clarify the legitimacy, or otherwise, of directed-energy weapons, high-power microwaves, cyber capability, robotic mechanisms, and even chemical sprayers. Commercial companies might also have to be involved.

There was an attempt to ban ASATs in 2014. Russia and China were keen on a much-revised draft text because it only banned space-based ASATs but allowed development and stockpiling of ground-based weapons. The US opposed it for those reasons, and little substantial progress has been made since. However, in 2022, the US took the lead and became the first country to announce a voluntary moratorium on "destructive, direct-ascent anti-satellite missile testing." Vice President Kamala Harris described these tests as "irresponsible" and said they "put in danger so much of what we do in space." However, the word "destructive" does give the US room to conduct computer testing and missile firing that does not impact targets.

For the immediate future, it looks as though we will only add to the space debris issue and must find ways to manage it.

So, can we simply shoot the debris out of the sky? One hurdle is that any machines intended to deal with space debris could have a dual purpose. In the near future, those firing directed energy—either at small bits of debris to disperse them or pushing larger bits into the atmosphere to be burned up—could also be used as weapons to attack spacecraft or satellites. Larger debris, such as defunct satellites, can be removed by spacecraft, but again governments worry that such craft could be used as cover for positioning hostile forces.

There are other ways to mitigate the problems with debris. We could introduce a globally agreed Space Situational Awareness system that catalogues all satellites, knows their directional capabilities, and then tracks them. All satellites could have small booster rockets fitted, allowing them to maneuver to avoid collision, and fall out of orbit relatively soon after their operational lives come to an end. Private companies are also working to secure lucrative contracts to build spacecraft capable of capturing the largest pieces of metal using nets and harpoons. But suppose a Japanese space trash-clearing company won a contract from the US to remove Chinese debris, including dead satellites, to make way for an American project?

Nature does some of the waste disposal job for us. Earth's gravity pulls trash into lower orbits and, if it's orbiting lower than 373 miles up, it normally falls into the atmosphere within a few years. Several hundred pieces take this route every year.

There are very rare examples of cars and houses being struck by meteorites or falling space debris, but only two documented cases of a human being hit. The first was Ann Hodges from Alabama. In November 1954, thirty-four-year-old Ann's afternoon nap was ruined by an eight-and-a-half-pound meteorite, which crashed through her ceiling, bounced off a radio, and hit her in the hip, leaving her with a huge bruise and a unique claim to fame. Then in January 1997, Lottie Williams was hit on the shoulder by a light piece of woven metallic material later found to have been from a Delta II rocket, which had reentered the atmosphere the night before.

As our planet is mostly water, that's where objects that get through the atmosphere usually land. Most debris burns up long before that, though. Coming in from the vacuum of space at thousands of miles an hour it is usually torn apart as it heats up because of the friction generated by colliding with the atmosphere. Some satellites are designed to break up easily on reentry in a process called Design for Demise. Most will disintegrate at about fifty miles above the ground and the pieces will be destroyed.

Objects can reach temperatures up to 3,000 degrees Fahrenheit, but the heat varies according to size, shape, composition, and angle of entry. Spacecraft are made to be streamlined and come in at an angle designed to create minimum friction. Space debris, though, is all shapes and sizes and comes in at an uncontrolled gradient. The pieces that survive are often those that have come in at a less destructive angle, and by chance are aerodynamically shaped. They are usually metals with high melting points, such as titanium, which, although it melts at about 3,000 degrees Fahrenheit, can make it through in the right conditions. The greater danger, however, comes not from debris hitting Earth but from its effects in space.

Professor Dolman lists the myriad problems facing those trying to develop safety measures: "The awareness system would need to be at a resolution only possible from sensors in space. Who would get to see the raw data? What else could it be used for? What are the potential military benefits? The second major issue is who would enforce compliance? What capabilities would the enforcer accrue? Who pays? Whose spacecraft would be used? Who would get the lucrative contracts to build, operate and maintain it?" Any plans and regulations relating to satellites become inseparable from military and national security issues.

It might seem sensible to set out "safe zones" for satellites in which other countries' machines cannot operate, but that conflicts with the concept of "innocent passage" and freedom of navigation rights we have worked out for the sea lanes on Earth under UNCLOS. It would also complicate future agreements to allow one state to inspect another's satellites to ensure they did not have dual use (civilian and military capability).

For the foreseeable future, therefore, space debris will continue to pose a danger to crucial satellite networks and space stations, and to human life.

There are numerous other areas in which agreements are lacking. For example, a powerful solar flare hitting Earth is entirely plausible, and in the Internet Age would have a massive immediate effect that would go on to wreck the global economy. Satellites in low Earth orbit as well as communication devices on Earth would be destroyed—the "internet apocalypse," causing blackouts, riots, and supply-chain disruption, as well as ruining your last-second eBay bid.

A small-scale version of this actually happened relatively recently. In March 1989, astronomers noticed a huge explosion on the Sun. Within minutes a billion-ton cloud of gas was heading toward Earth at over a million miles an hour. Two days later the cloud of electrically charged particles hit our magnetic field, creating electrical currents beneath North America. At 2:44 a.m. the following day, one found a weakness in Québec's power grid and two minutes later every light in the province went off. So did every computer, fridge, oven, elevator, traffic signal, and everything else needing electricity. In space several satellites were hit and spun out of control. It was twelve hours before the power came back on.

Given that our basic infrastructures rely on satellites, as do our commerce and militaries, what are nation-states doing collectively to protect them? Sangeetha Abdu Jyothi, a computer science specialist at the University of California, Irvine, has the answer: "To the best of my knowledge, there are no global agreements or plans to deal with a large-scale solar storm. A recent study estimates the economic loss during a catastrophic event in the US alone to be $40 billion per day. A solar storm will also affect every walk of life. Despite this, we lack a disaster preparedness plan for the worst-case solar events." On the plus side she says that extensive work is being done in the power grid sector to evaluate just how bad things could be and, because some regions are more at risk, studies are being done to see if low-risk countries could quickly launch new satellites to reestablish connectivity.

The same goes for the dangers posed by Earth being in the wrong place at the wrong time and getting hit by an asteroid bigger than 3,280 feet across. As the American science educator Bill Nye puts it, depending on the size, that would be "Game over. It's control-alt-delete for civilization."

We don't have an international plan for what to do in the unlikely event that the film *Don't Look Up* comes true. But it's not all negative. NASA, with international colleagues, has developed what it calls DART—Double Asteroid Redirection Test—to see if large objects on a collision course with Earth can be hit by a missile and deflected.

Its first test run was launched in November 2021 atop a SpaceX Falcon 9 rocket. The DART spacecraft, about the size of a large refrigerator-freezer, took ten months to reach a near-Earth asteroid named Dimorphos, which measures 525 feet across and orbits a bigger asteroid named Didymos. It hit Dimorphos head-on at 14,764 mph, causing it to slightly alter course and shorten its then twelve-hour orbit of Didymos by thirty-two minutes. It was a watershed moment, the first time humans had changed the orbit of a planetary object—and at around $314 million, it was money well spent.

Tackling such problems would be easier if laws existed to encourage cooperation between the major spacefaring nations, particularly the US and China. Expecting the world's two greatest powers to put aside their differences is naive, but if they can accept them and see past their mutual suspicions, both would benefit enormously from exchanging scientific expertise, as would the rest of the planet. China is already progressing with plans to develop an asteroid deflection system to help guard Earth from any city-sized lumps of rock heading directly our way.

Technology with which to spot approaching objects has developed to the extent that we can identify them more than twenty-five years in advance. There's an asteroid as big as the Empire State Building named Apophis that was first seen in 2004 and quickly identified as having a 2.7 percent chance of hitting us in 2029. Happily, this was later reassessed as having a 100 percent chance of a near miss, as it will pass within 20,000 miles of Earth. That's close, though. It will still take place in 2029—April 13, in case you want to mark the date in your diary. You can then add 2068, because scientists theorize that the effect of the 2029 near miss means Apophis may swing around again in that year. Luckily, they have updated research that says it will not hit us.

As well as the security fears that hinder agreement in such areas, there is also the issue of government budgets, especially in the democracies where such things are open to scrutiny. The problem is what

Everett Dolman calls the "Katrina Syndrome." Hurricane Katrina hit New Orleans in 2005, causing over 1,800 fatalities. It was called a "one-hundred-year" hurricane—something that might happen only once a century. Convincing voters to support paying higher taxes to protect against a possible "hundred-year" event is hard enough, but a "ten-thousand-year" event coming from deep space . . . who wants to campaign on that platform?

Nevertheless, the alarm bells have been ringing long enough for the awareness among scientists, space experts, war strategists, and environmentalists to begin to get through to politicians.

⸻

None of the things discussed here may happen, but without an adequate legal framework the temptation to try to make them a reality grows, especially when one country fears another may be gaining an advantage. We are already in an arms race in space, and it needs to be halted. Too often people fall back on the agreements of another age, notably the Outer Space Treaty.

We need greater clarity and shared commitments to transparency, pooled resources, debris collection, spacecraft disposal, freedom of navigation, deconfliction, release of data, situational awareness, and space traffic management, all within a respected rules-based order to which all parties agree. The Big Three space powers—China, the US, and Russia—agree on very little at the moment, and they know that what happens in space is an extension of what happens on Earth. They are ambitious and suspicious of one another's intentions—the Chinese and Americans both want to write the new international rules for space. They will need to be persuaded by everyone else that it is in their interests to cooperate.

The legal systems we have in place regarding space are nowhere near as comprehensive as in other spheres, such as maritime law. They require drastic updating, and in some cases need to be scrapped and new ones drawn up. Technology has outpaced law. Without laws, geopolitics—and now astropolitics—is a jungle.

5

CHINA:
THE LONG MARCH . . . INTO SPACE

The first to arrive is the first to succeed.
—Chinese proverb

It's 2061. Earth's surface is frozen solid. To escape the expanding Sun, the planet has gone on a journey. It no longer rotates because thousands of fusion-powered engines on one side of Earth propel it across our solar system. The farther from the Sun it travels, the colder it gets. Half the population is dead, and survivors live in vast underground cities. But Earth must reach Alpha Centauri, where there is a perfectly good, non-expanding sun that will allow us to get back to normal. "A journey of 4.3 light-years begins with a single step," as Confucius never said.

That's the plot of the thoroughly bonkers and entirely enjoyable 2019 Chinese sci-fi movie *The Wandering Earth*. On release it was a smash hit domestically, breaking box-office records. It was shown internationally on Netflix, and has become the fifth highest-grossing non-English-language film of all time. It is interesting on a number of levels, not least for what it says about soft power and how China projects its view of space.

Depiction of the Chinese
Tiangong space station

The director, Frant Gwo, says that in the US there is a narrative of humanity eventually leaving Earth to colonize the "endless frontier," and this is portrayed in American science fiction literature and films. But, he argues, the Chinese narrative is to improve life on Earth by using space resources. This is one of the themes in *The Wandering Earth*. Gwo told *The Hollywood Reporter*: "When the Earth experiences this kind of crisis in Hollywood films, the hero always ventures out into space to find a new home, which is a very American approach—adventure, individualism, but in my film, we work as a team to take the whole Earth with us. This comes from Chinese cultural values—homeland, history and continuity."

It's no surprise that it fits the messaging of the Chinese Communist Party (CCP), and that the party has supported the film. Gwo's movie was partially produced by the state-owned China Film Group Corporation and, as is normal in China, had to be approved by the publicity department of the CCP. The Ministry of Education recommended that it be shown in schools around the country. The party's Central Commission for Discipline Inspection felt moved to praise it, and the Foreign Ministry in Beijing did its bit for publicity, with spokeswoman Hua Chunying telling journalists, "I know the hottest movie now is *The Wandering Earth*. I don't know if you have watched or not. I'd recommend it." Fair enough, except that no one had asked her about the film.

Also fair enough is that, although there is a united Earth government in the film, it is a Chinese-led plan and Chinese heroes who save the planet, albeit with some help from a friendly Russian cosmonaut. It makes a change from Americans doing so, usually while saying things such as "I'm too old for this shit!" The rough equivalent in *The Wandering Earth* comes when a soldier fires an unfeasibly large machine gun at Jupiter while shouting, "Screw you, damn Jupiter!" This line probably tells you whether you wish to see the film or not.

The CCP leadership definitely wants you to see the film, however, because *The Wandering Earth* chimes nicely with "Xi Jinping Thought." Beijing knows that its growing space capabilities are viewed as a threat by the US and other countries. Using soft power, such as cinema, allows it to suggest to foreign audiences that there is nothing to fear from its activities, while simultaneously boosting domestic national pride and interest.

The Chinese president has long pushed the idea that China's space program is no threat to anyone; indeed, it seeks to work within international frameworks and for the good of humankind. So is its space program, entirely and directly controlled by the People's Liberation Army (PLA), for the benefit of all humanity?

It isn't—but then again, nor is that of any other country. However, China's space program is more militarized than that of any other state.

The National Space Administration comes under the State Administration of Science, Technology and Industry for National Defense. Its website says it was established "to strengthen military forces" and "serves the needs of national defense, military forces, national economy, and military-related organizations." The launch sites for rockets are run directly by the PLA via its Strategic Support Force, which is responsible for space, cyber, and electronic warfare missions. The department in charge of astronauts, or taikonauts as they are called, comes under the Central Military Commission Equipment Development Department.

None of this is confidential, but China appears keen not to advertise it. Government websites in Chinese are open about the military's control, including publishing pictures of senior officials in uniform, but their English-language versions make little mention of it.

Xi Jinping believes that China should have more of a leadership role in the world, and China sees space as an integral part of its plans. It takes a "techno-nationalist" approach to modernization, understanding fully that it needs to be a technological leader if it is to achieve its aims.

In the 1950s, Chairman Mao had been thinking along similar lines to Xi. He lamented that China couldn't even launch a potato into space. No one dared ask why he might want such a thing, and in the late 1950s, despite it still being a poor and primarily agricultural country, the decision was made to invest in long-range missiles and space technology.

China's version of America's Wernher von Braun and Russia's Sergei Korolev was Qian Xuesen (1911–2009), one of the greatest scientists the country has produced, who is regarded as "the Father of Chinese Rocketry." He graduated top of his class from National Chiao Tung University (now Shanghai Jiao Tong University), went on to study at the Massachusetts Institute of Technology, and then at the California Institute of Technology, where he stayed for nearly two decades. There, under the tutelage of Professor Theodore von Kármán, he was part of a team nicknamed "the Suicide Squad," due to their attempts to build a rocket on campus, and the ensuing accidents involving volatile chemicals.

During the Second World War Qian worked on America's response to Germany's V-1 and V-2 rockets, and on the Manhattan Project, which developed the first atomic bomb. After receiving the temporary rank of lieutenant colonel, Qian was sent to Germany to interview the V-rocket scientists, including von Braun. By the war's end he was considered one of the world's foremost experts on jet propulsion.

All this counted for nothing in 1949 when, as the Communist Party was seizing control of China, Qian was accused by the Americans of being a communist sympathizer. In 1950, he was stripped of his security clearance and put under house arrest. His subsequent application to return to China was denied by US authorities because he was the man who knew too much. When he was finally allowed to leave in 1955, he departed for China, telling reporters he would never set foot in the US again. He kept his word. America's loss—and China's gain.

While the communists were solidifying their control of China in the mid-twentieth century, they watched the Americans and Soviets spend billions in the Space Race. The bragging rights of who won were of less concern to the Chinese than the technological advances. The bigger and more far-reaching the rockets, the more alarmed Beijing became that they could be militarized and used against China. So Qian was set to work training a generation of scientists who helped develop China's nuclear bomb and the country's Dongfeng ballistic missile system.

In 1956, in the spirit of "brotherly assistance," the Soviets provided Qian with the blueprints for their R-1 rockets and sent specialists to Beijing to jump-start the Chinese missile program. A testing site was built in the Gobi Desert and dozens of Chinese students were sent to Moscow for training.

The Chinese wanted access to more modern rockets, but there were limits to "brotherly assistance" and the Russians were reluctant to allow their latest technology to be transferred to another country. The Chinese students resorted to copying restricted documents and tapping their instructors for knowledge.

Relations between Moscow and Beijing were deteriorating on a range of issues, including a border dispute in the far east and the fact that both claimed leadership of the Communist world, each insisting that its version of Marxist-Leninism was the correct form of communism. Chairman Mao also felt that the Soviet leader, Nikita Khrushchev, was not aggressive enough against the "running yellow dog capitalist" nations of the West.

By 1960, cooperation was withdrawn. But building on what they knew, the Chinese were able to produce the Dongfeng, or "East Wind," class of missile with capability for short, medium, intermediate, and eventually intercontinental ranges, which could be fired from silos or mobile launchers. Qian used this rapid absorption of technical knowledge to oversee the launch of China's first satellite and laid the foundations for the Chinese space program.

Qian is a national hero and there is an entire museum containing seventy thousand artifacts dedicated to him. His story is a warning about rejecting outside scientific knowledge based on flimsy suspicions of intent. Former US Navy secretary Dan Kimball said America's treatment of Qian was "the stupidest thing this country ever did."

In 1967, Mao gave the order to put a taikonaut into space and the first candidates were chosen for training. But the program was canceled as the country was engulfed in the chaos of the Cultural Revolution, during which many scientists were imprisoned or killed. For example, Zhao Jiuzhang, head of China's satellite program, was denounced as a "counterrevolutionary" and beaten up by the Red Guards. He is thought to have drowned himself in a Beijing lake.

Despite these setbacks, China's first satellite was delivered into orbit on April 24, 1970. It circled the globe for twenty-eight days. China thus became the fifth country to send satellites into orbit following the Soviet Union, the US, France, and Japan. Five batteries inside the machine were used to allow the song "The East Is Red" to be beamed back to Earth so that we all could enjoy the lyrics (repeated every thirty seconds): "The East is red. The Sun has risen. China has birthed a Mao Zedong!" In China, April 24 is now "Space Day."

From there, the program moved quickly. By the mid-1980s, China was launching satellites on a regular basis and offering its facilities to other countries.

For the first few decades, China's space program was primarily about achieving military ambitions, in addition to using satellites for monitoring the weather and, as the country began to industrialize, deciding where to put roads and railways. In this century, though, the Communist Party has understood its use in persuading everyone to understand China's place in the world—that is, as being among its military, technological, and economic leaders with the potential to become the preeminent power.

When China deliberately destroyed its own weather satellite using a KKV in 2007, other countries were horrified about the ensuing space debris but impressed—and alarmed—that the Chinese had pulled off the equivalent of hitting a bullet with a bullet: traveling at about 18,020 mph, with just a second to go before impact, the KKV made three lightning-quick adjustments to its trajectory to hit the 6.5-foot-long satellite square on.

China said it wasn't part of an arms race in space because it would never become involved in such a thing. If so, then the allegations that Beijing is powering ahead with research on ground-based directed-energy weapons to hit enemy targets in space are baseless. Beijing's version also means that the remote sites in China with large buildings, which have roofs that slide back and domes that could be deployed for targeting, are probably only used by enthusiastic astronomers. Other uses might be imagined.

In early 2022, Beijing published a "Perspective" on its space program, beginning with a quote from President Xi Jinping: "To explore the vast cosmos, develop the space industry and build China into a space power is our eternal dream." Throughout the document are references to how the space industry will contribute to China's growth and "to global consensus and common effort with regard to outer space exploration and utilization, and to human progress." The document goes on, and on, to list China's achievements to date amid statements of intent. There are plans for next-generation manned spacecraft, a human lunar landing, an international research station on the Moon, probing asteroids, and deep space exploration. There's also a sentence about "exploration of the Jupiter system, and so forth," which is intriguing, but the "and so forth" may have lost something in translation.

Its "Vision" is to "freely access, efficiently use, and effectively manage space." The "freely access" and "effectively manage" parts are a shot across the bow of the Americans and any attempt to deny China its place in the heavens. In 2017, the head of its Lunar Exploration Program, Ye Peijian, said: "If we don't go there now even though we are capable of doing so, then we will be blamed by our descendants. If others go there, then they will take over, and you won't be able to go even if you want to. This is reason enough."

The document is explicit in its call for the United Nations to have the central role "in managing outer space affairs." It points out that since 2016, China has signed space agreements or memoranda of understanding with nineteen countries and regions, including Pakistan, Saudi Arabia, Argentina, South Africa, and Thailand, and four international organizations. It emphasizes cooperation with the European Space Agency, Sweden, Germany, and the Netherlands. It trumpets that it has provided satellite launches for a range of countries and opened its facilities to developing states such as Laos and Myanmar.

This is all pushback against what Beijing sees as the US trying to dominate governance of space. Over the years, there have been efforts to collaborate with America. In early 1984, President Reagan offered a place on the US Space Shuttle to a taikonaut. In 1986, a group of Chinese scientists were due to visit Houston's Johnson Space Center as part of the preparations, but that January the *Challenger* exploded seventy-three seconds after launch, killing all seven crew members. The visit was canceled and all "guest programs" suspended indefinitely.

Now, China has been frozen out of the Artemis Accords after Congress limited NASA's ability to cooperate with it under 2011's Wolf Amendment. The rationale of then Republican congressman Frank Wolf was that the relationship between space exploration, technological advances, and China's military was such that the US could not risk collaboration with its growing rival. Specifically, the concern was about the possibility of intellectual property theft from NASA computers and any joint US-Chinese research, which Beijing could apply to sensitive military technologies, including ballistic missiles.

Chinese hackers are known to have briefly gained entry to computer systems at the Department of Defense, the Office of the Secretary of Defense, the U.S. Naval War College, the Pentagon, a nuclear weapons lab, and the White House. More traditional espionage has also been uncovered. For example, in 2008 Shu Quan-Sheng, an American physicist living in Virginia, was convicted of transferring information about the liquid hydrogen tanks of a US rocket to Beijing. In 2010, Dongfan Chung, a former engineer at Boeing, was convicted of providing China with more than 300,000 pages of sensitive information, including data about the US Space Shuttle.

China has reacted to being frozen out by constructing a rival to the International Space Station, forming strategic scientific relationships with a host of countries, and building a domestic space industry that looks at least as cutting edge as that of the United States. This has been achieved without input or monitoring from the Americans.

Impressive. And quick, given that the first Chinese person went into space in 2003: thirty-eight-year-old military pilot Lieutenant Colonel Yang Liwei. His capsule was propelled into orbit by one of the Chinese-developed rockets, the Chang Zheng 2F. Yang Liwei orbited Earth fourteen times during a flight of twenty-one hours—and China became the third country to send a human into space. *China Daily* called it the "Great Leap Skyward."

The achievements kept coming. The year 2012 saw the first Chinese woman in space—fighter pilot Major Liu Yang. In 2014, China completed its coastal spaceport at Wenchang, specifically for the larger-diameter Chang Zheng rockets, which need to be launched over water. In 2016, two taikonauts spent a month on board China's Tiangong-2 space station after their craft successfully docked with it.

In 2019, the uncrewed Chang'e 4 became the first spacecraft to land on the far side of the Moon. The mission was another example of the potential for cooperation between the Chinese and Americans. NASA was given dispensation to help provide information for the landing site, and later the two countries agreed that findings from the coordinated activity could be shared with the international research community via the United Nations. Another notable moment came in 2020 when the final BeiDou satellite was put in position, completing a navigation network to challenge the American-owned GPS. "*Beidou*" is the Chinese word for the Plough constellation, or the Big Dipper. Then 2021 saw the first spacewalk by a Chinese woman, Wang Yaping.

Perhaps the biggest milestone in the last decade was orbiting, landing, and then deploying a rover on Mars. The Tianwen-1 mission arrived at the planet in February 2021 and spent three months searching for the right location to touch down. On May 14, the lander left the orbiting vehicle and made a soft landing. The Zhurong (god of fire) rover was then released to conduct surveys of Martian geology, search for water, and beam back sound and vision. There are now three active rovers on the planet: the Zhurong, and the Perseverance and Curiosity from two earlier NASA missions.

All of this is a source of great pride in China and is interwoven with the mythology of the Communist Party. China's Chang Zheng ("Long March") rockets are named after a famous military retreat during the Chinese Civil War in 1934–35, in which the Red Army covered 5,600 miles over rugged terrain. It helped Mao come to power and go on to defeat the anti-Communist forces. It is part of the foundation myth of the Chinese Communist Party and is often used as an example of heroic sacrifice to achieve greatness. To use the term for the rockets propelling China to greatness in space is deeply symbolic.

Interestingly, however, in recent years China has softened some of its public exhortations about the superiority of communism and instead has embraced elements of nationalism and myth from an older historical collective memory. This is reflected in the naming of space missions and equipment. For example, in 2007 the uncrewed craft that orbited the Moon was called the Chang'e 1 after a beautiful woman, as the story is told, who stole the elixir of immortality from her husband, drank it, flew to the Moon, and became a celestial goddess. Chang'e had a pet rabbit called Yutu ("Jade Rabbit"), which now spends its time hopping about on the Moon pounding the elixir of immortality in a mortar to ensure Chang'e has enough of the stuff. So, it's no surprise that when China landed the Chang'e 3 on the Moon in 2013 the rover that trundled off it was called Yutu.

Meanwhile, on board Tiangong space station, taikonauts, who travel to it in a Shenzhou ("Divine Vessel") capsule, can thank their lucky stars they are in the "Heavenly Palace," named after the residence of the Celestial Ruler who holds supreme authority over the universe in Chinese mythology. The word "taikonaut" is a mixture of Mandarin and Greek, from *taikong*, meaning "cosmos," and *nautes*, Greek for sailor. The term was popularized by Chinese space analyst Chen Lan, who runs a website called Go Taikonauts! The official name for a Chinese astronaut is a *yuhangyuan*, or "traveler of the universe" (or, worse still, "universal travel worker").

These names are important. They signal to the world that space is not just the domain of Americans and Europeans, and that for every Artemis there is a Chang'e.

———

There were setbacks in the development of China's own space program (as with Russia's and America's). These include the tragedy of

1996, when a rocket exploded after take-off and killed at least six people on the ground. The exact details are still not known—a reminder that China remains in many ways a closed society. In 1972, the double Pulitzer Prize–winning historian Barbara Tuchman returned from a visit to China and wrote: "In addition to the language barrier, I am trying to analyze a relatively secret program managed by the government of one of the closest societies in the world. The fact that the People's Republic of China has embraced capitalist modes of production to make its economy grow faster should not obscure the fact that this is a communist nation, ruled by a communist party, where secrecy is government policy at all levels." China has changed in many ways, but Tuchman's words are as true now as then.

Despite the legacy of secrecy surrounding China's space program, it is now common knowledge that its launch capacities are well established—and expanding. The Chinese National Space Administration (CNSA) has several launch sites across the country. The oldest is the one near Jiuquan in the Gobi Desert, from where Yang Liwei took off in 2003. The desert is also home to the Taiyuan facility, which launches some of China's weather satellites but is also part of its intercontinental ballistic missile system. Sichuan Province hosts the Xichang Satellite Launch Center, and the more modern Wenchang Space Launch Site on the South China Sea island of Hainan is now used to get taikonauts to the Chinese space station and for longer uncrewed missions. A fifth launch facility is being completed in the eastern port city of Ningbo, about two and a half hours' drive down the coast from Shanghai. Within a few years it is expected to be launching one hundred commercial rockets a year in what are called "quick-fire" liftoffs. The Ningbo site has similarities to the Kennedy Space Center in Cape Canaveral, Florida. Both are on the coast, meaning rockets do not have to fly over land, and are on favorable latitudes for breaking out of the atmosphere quickly. Mission control is usually overseen either from Beijing or Xi'an in central China. There is also a global network of Chinese ground-tracking stations, which help keep an eye on space traffic and communicate with China's satellites and space station. They are based in a variety of countries, including Namibia, Pakistan, Kenya, Sweden, Venezuela, and Argentina. The CNSA also has a fleet of tracking ships dotting the oceans. They are an odd sight, their decks bristling with huge antennae dishes alongside banks of lasers as they roam the seas and sweep the skies, watching for satellites and missiles.

The new spaceport at Ningbo is only a few miles from a cluster of commercial launch industries near the mouth of the Yangtze River. With access to a huge port and other space-focused industries in Shanghai, Ningbo, unlike the other main launch centers, is well placed to integrate with existing supply chains and looks destined to play a key role in the future.

Local officials want Ningbo to be known as "China's Space City," although it has competition in that desire. The country's largest car manufacturer, Geely, has its headquarters there and is investing heavily in satellite design and aerospace-related industries. In 2022, it used the Xichang facility to launch nine of its own satellites into low Earth orbit as the first stage of a network to provide more accurate navigation for autonomous vehicles.

This is all part of the growing commercial space industry in China. It remains behind the US in terms of private funding, but companies are keen to invest, especially in designing, building, and launching satellites before low Earth orbit gets too crowded. The CCP began to encourage private investment in 2014; however, as with all Chinese enterprises, the link to the state is stronger than in most countries. There are now more than one hundred private space-related companies in China, but many are spin-offs from the government sector. For example, the rocket manufacturer ExPace, located in the Wuhan National Space Industry Base complex, is a subsidiary of the state-owned China Aerospace Science and Industry Corporation.

Others have more distance from the state: i-Space, for example, was the first private Chinese company to achieve orbit when it launched its Hyperbola-1 rocket in 2019. However, this was followed by two failures in 2021 and another in 2022. Others have also experienced serious setbacks. To help counter this, the government is slowly allowing the transfer of previously restricted state technology and expertise into the sector as part of a national strategy of civil–military fusion. This links the state, private enterprise, and the country's top research universities in clusters of technological excellence in a more formal manner than in the US. In a very competitive market, some of the new companies are bound to fail, but what is equally sure is that some will rise to become powerful national and, probably, global players.

In all this China will be helped by a huge, dynamic workforce. The country is predicted to have long-term demographic problems— a third of the population may be over sixty by 2050—but for now it can still turn out large numbers of scientists and engineers. Beihang University (Beijing University of Aeronautics and Astronautics) alone has 37,000 students. Each year this century China has increased the number of engineers who graduate, and there is no way given its population that the United States can compete in terms of volume.

In the near future, Beijing intends to further develop its BeiDou

Navigation Satellite System for use in various industries, having seen the boost GPS has given to the US economy since the mid-1980s: American farmers use it to plan the best use of their land, delivery services are steered more efficiently through cities, financial institutions can time-stamp transactions, and ship owners can track their fleets en route to ports. Studies suggest GPS has boosted the US economy by $1.4 trillion, with most of the growth coming in the last decade. The BeiDou system already transmits to more than 400 million cell phones and 8 million vehicles. Its encrypted military application is more accurate than the civilian version and will be used to monitor the movements of the PLA and the armed forces of other countries.

China also intends to launch at least a thousand satellites over the next decade and may well launch many more than that. It will increasingly offer its services to developing countries that cannot afford to launch rockets or have satellites of their own. This will be used to cement bilateral ties in an attempt to pull nations away from the US. Satellites used for scientific discoveries are likely to have some notable triumphs to rival that of the Hard X-ray Modulation Telescope, China's first X-ray astronomy satellite, which observes black holes and has discovered the strongest magnetic field in the universe.

There may already be a Chinese spaceplane in service. If not, they will build one. A spaceplane is a rocket with wings that takes off vertically, gets itself into space up to 497 miles above Earth, can maneuver, and lands like a plane. The Americans have had one since 2010: the X-37B resembles the now-retired Space Shuttles but is smaller, about 30 feet long. It's only flown a handful of missions. What they were is secret.

Even less is known about the Chinese version—the Shenlong or

"Divine Dragon." It's thought to have flown into space at least once, but even that's not certain. What is on the record is the ambition to land on passing asteroids and mine them for the riches therein. Some of these rocks are tens of miles wide and contain billions of dollars' worth of the metals required for twenty-first-century technology. One of China's many start-up companies, Origin Space, has already launched a robot prototype to capture and destroy space debris and intends to develop it to be able to mine asteroids.

There's also the intention to send another probe to Mars. Just getting there is hard enough, but China, along with the US and the European Space Agency, is working on plans to dig up some soil and rock samples and get them back to Earth. Farther out, they want to send probes to Jupiter and Saturn.

But perhaps the project with the greatest political significance is China's coming Moon landings.

In 2021, China and Russia signed a memorandum of understanding that they will jointly build a base on the Moon called the International Lunar Research Station (ILRS). They envisage three phases: first, reconnaissance up to 2026, including three crewed missions; then landing on the Moon, followed by "returning." The "returning" is envisaged as a Moon base that will be habitable by 2036. A statement from the Chinese said the two countries would "conduct scientific exploration at the lunar south pole in order to facilitate the construction of a basic structure for a lunar research station in the region." The south pole has been earmarked because its icy craters are a potential source of water.

When China landed an uncrewed craft on the far side of the Moon, it planted the Chinese flag and began digging for rocks in a region it is considering using as a base. Some reports suggest China wants a permanent presence on the Moon as early as 2028, but this seems beyond ambitious; 2030 is more realistic, and even that would be impressive.

The first structure built will enable mining, to extract the resources that will allow the base to grow—central to this is water, hence the south pole. Moscow and Beijing say they intend the base to be fully open by 2035; the American-led Artemis project is more vague about its timetable.

Building a base on the Moon will capture a generation's imagination in the way the 1969 Moon landing did. Flowing from that will be an appreciation of the technological brilliance and, equally importantly, resolve of the nation or nations that do it first. This is not just about "planting a flag," it's about seizing the "High Frontier" for both military and commercial advantage. The prizes are the Moon's potential riches, and the ability to use it as a gravitational point to deploy military satellites that would be difficult for competitors to detect.

Further claims on the "geography" of space will be made as the decade progresses. China is already the only country operating its own space station, the Tiangong-3. It doesn't make headlines in the way a Moon base will, but in astropolitical terms, having the only sovereign station is quite a statement. The better-known ISS is a "cooperative program" involving European countries, Japan, Russia, the US, and Canada, and has hosted 266 astronauts from twenty different countries. However, the Tiangong is owned and operated solely by China and is expected to be in service up to about 2037. Versions 1 and 2, built and launched between 2009 and 2016, were test versions for the third, which is almost three times heavier and much bigger. Deputy chief designer Bai Linhou says the three taikonauts on six-month missions will feel as though they are "living in a villa." If so, it will be more akin to a dodgy Airbnb rental than an all-mod-cons dream holiday getaway. It has only three modules, whereas the ISS has sixteen. Still, the views are good and will get better once the Xuntian Space Telescope joins it. Although it is a similar size to the Hubble Telescope, with a 6.6-foot-diameter mirror, Xuntian is said to have a field of view three hundred times greater and a camera that is capable of capturing 2.5 billion pixels. On board the Tiangong the taikonauts are researching space medicine, biotechnology, microgravity combustion, fluid physics, 3D printing, robotics, directed-energy beams, and artificial intelligence. The station is usually about 249 miles up and, like the ISS, sometimes visible to the naked eye.

The ISS is marked for decommissioning by 2031 at the latest. As it closes, a small window may open for China. The Artemis program includes the Gateway—a small station orbiting the Moon acting as a hub to allow spaceships, crews, landing modules, and rovers to resupply during frequent trips (as we'll see in the next chapter). But any serious delays in building the Gateway will leave the Tiangong as the only place open for guests, and that will be a demonstration of Chinese hospitality, spirit of cooperation, and . . . leadership.

Beijing has already said it hopes to host visits by international astronauts and wants to work "with all countries in the world committed to the peaceful use of outer space." It has approved a range of scientific experiments to be conducted on board the station from a list of forty-nine submitted by various countries. For example, a Norwegian-led research program called Tumors in Space was chosen. From 2025, it is scheduled to look at how the microgravity and radiation found in space affects the growth of tumors.

China and the US look destined to spend the next decade mostly isolated from each other regarding high-tech science and engineering, two areas that are crucial in addressing the challenges humanity faces both on Earth and in that most hostile of environments for humans, space. Cooperation is possible. Replacing the Wolf Amendment with something less like a sledgehammer would help, and even as that is worked on, because the amendment is specific to NASA, the US Departments of Defense and State have room to explore bilateral avenues of mutual benefit.

Détente between the Americans and Soviets was helped by the Soyuz–Apollo "handshake in space." Following the end of the Cold War, collaboration between Russia and the US on the ISS was a bridge on which at least to try to build a better relationship. The return to the Moon is another opportunity.

Whether either side is able or willing to make that leap in space depends on their relationship on Earth.

6

THE USA: BACK TO THE FUTURE

When men are arrived at the goal, they should not turn back.
—Plutarch

Been there, done that—and a million people bought the NASA T-shirt. So why return?

The last time humans were on the Moon was more than half a century ago: December 14, 1972, when Eugene "Gene" Cernan and Harrison Schmitt became the eleventh and twelfth people to walk on its surface. The question about whether to go back has been asked by Americans ever since.

Within the debate are a variety of answers. There are those who take the view that space exploration is simply too expensive, and that humanity's focus must be on more earthly problems. Others argue that we should be shooting for Mars and the priority is just to get straight there. At the moment the argument has been won by those who say we must get back to the Moon for a variety of reasons, among them because the Moon is the staging post to the Red Planet; the intention is to be back on the lunar surface well before the end of this decade.

The SLS Exploration Upper Stage with the Orion spacecraft, part of NASA's program for returning to the Moon and for deep space exploration.

Significant in its absence is a similar debate in China. There, it is taken as a given that space exploration is a vital part of national development. There is a clarity of purpose exemplified by President Xi Jinping's declaration that "the space dream is part of the dream to make China stronger."

Naturally, the Politburo in Beijing is unencumbered by such irritations as opinion polls, opposition political parties, and democratic budget oversight. As such, the Chinese space program is stable. The US equivalent? Not so much.

Space continues to capture the imagination of huge swathes of the American public, but as a policy it barely figures in elections and so can easily be diverted to the backwaters of budget planning. It is periodically buffeted by political whims and economic headwinds. At times it is in fashion and used as inspiration, at others it is considered a costly headache.

This was especially true in the years after the Moon landings. American technology was triumphant and the Space Race had been won. Public interest fell away, so did funding. As the American writer Tom Wolfe memorably put it, the Moon landing ended up as "a small step for Neil Armstrong, a giant leap for mankind and a real knee in the groin for NASA."

In President Kennedy's 1961 speech vowing to put a man on the Moon by the end of the decade the optimism and drive of early 1960s America rang out. When it comes to the US space program, nothing since has matched this confidence and understanding of the relationship between space and geopolitics, although Ronald Reagan's administration came close. Kennedy's rhetoric was very much of its time, and that time was the Cold War.

All of the crewed Moon landings took place during Richard

Nixon's presidency (1969–74), but he had inherited the Apollo project from his predecessors. NASA had drawn up ambitious plans to build a Moon base by 1980 and send astronauts to Mars by 1983, but Nixon canceled them in favor of the Space Shuttle, which went into service in 1981. He'd called the Apollo 11 mission "the greatest week in the history of the world since the Creation," but within a few months of the triumph he was telling aides that he did not see the need for American astronauts to keep returning to the Moon. He was conscious of the cost and dangers involved in the Apollo missions, and also knew that public interest was waning following the first lunar landing.

So it was in 1972 that Apollo 17's Harrison "Jack" Schmitt and Gene Cernan piloted the final crewed flight to the Moon. As Cernan took the last few footsteps back to the landing module he paused and knelt to write his daughter Tracy's initials in the dust—TDC. Then he made a short speech: "We leave as we came and, God willing, as we shall return: with peace and hope for all mankind." The hatch closed, his fingers hovered on the craft's ignition button and, as he later recalled, the last words spoken from the Moon were uttered: "OK, Jack, let's get this mutha outta here."

It was a strange end to a project that can be considered humanity's greatest scientific and technical achievement. When the landing module docked with the main spacecraft the crew held a live press conference. The US networks didn't bother to air it.

The Moon was history, and expensive history at that. NASA needed a less costly alternative to a one-shot disposable rocket, and a project it could justify to the White House. The reusable Space Shuttle was intended to provide the US with a low-cost way to get people and payloads into low Earth orbit. It managed the latter, but at a financial cost far above the original budgetary projections, and a cost to human life that revealed technical design flaws.

The first orbital test flight came in 1981 and over the next thirty years the Shuttle Program flew 135 missions. Among many achievements shuttles docked with the Mir space station, carried the Hubble Space Telescope into orbit, and helped build the ISS. However, the explosion of the *Challenger* in January 1986 was a disaster for the program. In his address to the nation President Reagan paid tribute to the crew: "The future doesn't belong to the fainthearted; it belongs to the brave. The *Challenger* crew was pulling us into the future, and we'll continue to follow them." An investigation suggested that NASA officials had made too many assumptions that the shuttle could survive minor faults on launch and the program was grounded for almost three years before returning with numerous design changes to the rocket boosters used in the launch process.

On the military front Reagan supported "Star Wars"—the Strategic Defense Initiative—which proposed a network of missiles and lasers in space and on the ground. It was never built due to numerous technical difficulties in developing the lasers and to political opposition on the grounds that it would cause an arms race with the Soviet Union. However, some of the work done on the technology paved the way for current missile defense technology.

President George H. W. Bush (1989–93) was a supporter of building bases on the Moon and Mars but could never persuade Congress to fund development. His successor, Bill Clinton (1993–2001), was in charge at a time of economic growth. The construction of the ISS began halfway through his second term in office, but there was little talk of the Moon or beyond.

That changed when George H. W.'s son George W. became president (2001–09). In 2003, there was a second shuttle disaster when the space shuttle *Columbia* broke apart while reentering Earth's atmosphere, again with the loss of all seven crew. Following its initial test flight, the shuttle now had a ratio of one fatal crash per sixty-seven flights. NASA had said that shuttles could launch every month, but in reality it struggled to fly more than one every three months and at a cost that led commercial companies to look for alternatives to get their satellites into orbit. The following year Bush outlined a plan to retire the entire fleet and concentrate on a return to the Moon by 2020.

NASA was given funds to develop a more modern crewed spacecraft, a lunar lander, and two new rockets. The then director of NASA, Michael Griffin, described the plans as "Apollo on steroids." It was not to be. There were delays and cost overruns as NASA went over budget by $3.1 billion. In 2009 came President Barack Obama, very much of the "been there, done that" view. One of his first acts was to cut the funding. Instead, he said America should aim for a different target—asteroids—and move on to Mars. Not a great deal happened, and then the Donald arrived.

President Obama had torn up President Bush's plans; now President Trump tore up those of Obama. Asteroids were out. The Moon was back in fashion. It wasn't just that Trump seemed keen to reverse most of the Obama administration's work; space travel was becoming cheaper, technology had advanced, there might be water and precious metals on the Moon, and Beijing looked as if it intended to make one giant leap for China.

The Artemis program Trump announced in 2017 intends to put men and women on the Moon this decade and have a base there in the 2030s, before eventually heading out to Mars. US taxpayers are expected to pay $93 billion toward the project and that's only up to 2025.

President Biden inherited the plan and gave oversight of the program to Vice President Kamala Harris. NASA is committed to its goals and the government is committed to its budget, but it is instructive that Biden pretty much ignored it as he was more focused on the military and commercial aspects of US space policy.

This is broadly in line with the public's priorities. In 1969, 53 percent of Americans felt that the benefits of US space policy were worth the financial costs, but by the mid-1970s only 40 percent did. Since the 1980s the figure has remained above 50 percent. A survey in 2021 by Morning Consult found that only 8 percent of respondents felt the government should not be investing in space, but 40 percent thought it was "not too important a priority" and 31 percent considered it "an important but lower priority." The same poll asked about the public's priorities for government involvement in space. Some 63 percent of people thought that the main issue should be to help combat climate change, while 62 percent believed that monitoring asteroids that could hit Earth was a major priority. However, only about a third prioritized sending astronauts to the Moon or Mars.

These figures are a reflection of priorities, not of a lack of interest in space. In many countries there is an acceptance that space travel is a matter for the state, but the US is a special case. Americans are more ready to argue that private industry should make the running and may even be better equipped to take on the incredible challenges presented by space travel. The effects of this attitude are clear. In the commercial sector, American companies are ahead. Investment and competition are rising as businesses weigh the costs and benefits of mining possibilities on the High Frontier.

It has also emerged in polling, however, that the majority of Americans believe China is a "major threat" to US leadership in space and want to maintain US dominance. Despite this, when it comes to building a base on the Moon there is not the same urgency to "win the Space Race" as there was in the Cold War. But on the military front, the US is determined to meet any challenge from China or Russia.

In the previous chapter we looked at the Chinese government's space policies and aims. Those of the US are remarkably similar. This is good and bad. Good in that both talk about cooperation—for example, in its 2022 Space Priorities Framework document the US says it will "demonstrate how space activities can be conducted in a responsible, peaceful, and sustainable manner." But it also says: "The United States will lead in strengthening global governance of space activities." Not according to China and Russia it won't.

The document doesn't name those countries, but it's difficult to see who else the following paragraph is aimed at: "The military doctrines of competitor nations identify space as critical to modern warfare and view the use of counterspace capabilities as a means both to reduce U.S. military effectiveness and to win future wars." Hence, "To deter aggression . . . the United States will accelerate its transition to a more resilient national security space posture . . ."

The tensions have been building for some time. Shortly after China's 2007 KKV satellite hit, China suspected that the US was sending a message back via a missile launch of its own. However, it's equally plausible that America's destruction of one of its own top secret spy satellites was not a response to the Chinese action.

At 10:26 p.m. on February 20, 2008, a missile fired from the USS *Lake Erie* took off for space. Four minutes later it hit satellite USA-193 at an altitude of 150 miles. This was no past-its-sell-by-date satellite, but a state-of-the-art machine bristling with the latest top secret spyware. Shortly after it had gone into orbit in December 2006, the Americans had lost control of the satellite, which was the size of a bus. The risks from debris if it fell to Earth were low, but the machine still held about a thousand pounds of highly toxic hydrazine fuel in a titanium tank, which has a high melting temperature. NASA briefed President George W. Bush that the potential casualty figures of an uncontrolled reentry of the satellite were the highest ever associated with such an event. He approved Operation Burnt Frost to shoot it down.

The challenge for the US Navy was to hit a target traveling faster, and at a higher altitude, than any engaged during the years of testing the *Lake Erie*'s Aegis Ballistic Missile Defense System. This was not a rehearsal. For the Americans it was uncharted territory. They were aiming for the satellite's fuel tank. A glancing blow might not be enough. The closing speed was moving past 21,748 mph before impact, leading to a massive detonation as the fuel exploded in a brilliant flash of light. Debris was thrown out, but a far smaller amount than that created the year before by the Chinese KKV.

Beijing and Moscow viewed Operation Burnt Frost as a continuation of America's Cold War military activity in space. The operation may not have been planned to bring the US into the modern ASAT era, but it did. Since then, US military capacity in space has increased every year.

In 2019, the government launched Space Force, the most recently formed of the six branches of the US Armed Forces (army, navy, marines, coast guard, and air force). It has a four-star general who, along with the other heads of the military, is a member of the Joint Chiefs of Staff. It is responsible for GPS satellites, which can detect missile launches, and has ground-based jammers that can block transmissions by enemy satellites. It also tracks space debris.

Its budget—around $26 billion annually—will probably grow in line with awareness of the centrality of space in modern warfare. It's currently the smallest of the armed forces, with just sixteen thousand military and civilian personnel serving in various locations around the country, including the headquarters in the Pentagon, Cheyenne Mountain in Colorado, and the air force base in Los Angeles. As a young organization it lacks a strong institutional culture but, conversely, as a "start-up" can benefit from new ideas. On a less important note, more thought should have gone into its logo. It so boldly resembles that of *Star Trek*'s Starfleet Command that George Takei (aka "Sulu"—one for older readers) was moved to comment, "We are expecting some royalties from this . . ." On the plus side, there's some nice alliteration going on with its motto: "Semper Supra," meaning "always above."

Since its inception there has been debate about its role. When Space Force was created some critics said it "militarized" space, but that misses the point that space has been militarized from the moment humanity first broke through the atmosphere. Space Force was built from units already doing similar work within the US Air Force and, as we saw earlier, the Soviet Union and the US were using satellites to spy on each other during the Cold War. The mantra "space is a warfighting domain" might be described as aggressive, but it is a statement of fact.

In practical terms, should Space Force be responsible for projecting military power all the way into deep space, or should it support the traditional warfighting services through surveillance, missile warning, communications, positioning, and navigation? The latter approach appears to be winning at present. Even though its name conjures up visions of US spaceplanes firing lasers at enemy bunkers on the Moon, it looks likely that the much bigger branches of the armed forces will win the inevitable turf war and keep control of the overtly offensive side of space war.

In military terms, the US has a clear lead on China—for now. In 2021, General David D. Thompson from Space Force warned: "The fact that, in essence, on average, they are building and fielding and updating their space capabilities at twice the rate we are means that very soon, if we don't start accelerating our development and delivery capabilities, they will exceed us." His time frame was 2030. General Thompson's forecast may turn out to be right, but China still has a long way to go to even get close to America's capabilities. The budget for Chinese military space activity is opaque, but almost certainly considerably less than that of the US. As of early 2023, there were roughly 4,900 active satellites in orbit; almost 3,000 of them are American and about 500 are Chinese.

Washington is investing heavily in early-warning satellites that use sensors to detect the infrared heat signatures coming from ballistic and hypersonic missiles. They then transmit the data securely to military command centers on Earth. These are part of the "Tracking Layer" of satellites the US is building in low Earth orbit. By 2028, it hopes to have one hundred of them acting as a shield against high-speed maneuverable missiles.

Money is also being spent on developing lasers that will eventually be deployed in space. The US Navy has had versions of a Laser Weapon System since 2014, but in 2022, its capabilities were shown to have advanced when the navy successfully used an all-electric, high-energy laser weapon to shoot down a fast-moving cruise missile. An invisible beam of energy homed in on the missile and after just a few seconds parts of it began to glow orange, then smoke poured from its engine and it tumbled downward. Once the system has been built, the actual "kill shot" costs just a few dollars in electricity. By comparison, a single guided missile can cost tens and even hundreds of thousands of dollars. As far as is known, the lasers are only deployed on Earth, but if any spacefaring country arms their satellites with them, others will follow suit.

One growth area will be "secret-not-so-secret" reusable spaceplanes. Space Force has control of the crewless X-37B, which spent more than two years in space on what is thought to have been its sixth mission. Most of what it was doing for so long is classified, and Space Force's beige statement that it's "an experimental test program to demonstrate technologies for a reliable, reusable, unmanned space test platform" is unlikely to assuage Chinese and Russian claims that it is a weapon. At one point in the operation, the head of a Russian defense company asserted that the plane was carrying three nuclear bombs that it could drop on Moscow from orbit.

The incompatibility of that idea with both physics and military tactics puts it somewhere on the scale between implausible and silly. Another claim that the X-37B was being used to spy on Russia is less far-fetched, but even then, it's difficult to see what the plane can do that satellites cannot. It's probable that there is a military aspect to X-37B, but hiding nuclear weapons in it before it blasts off, using thousands of gallons of rocket fuel, is unlikely to be it. I don't know what it does. But I want one.

Whatever is happening at Space Force, they're aiming high. In 2020, a document defined the geographical limits of its mission. The wording suggests there are none. It says: "Until now, the limits of that mission [defending and protecting US space interests] have been in near Earth, out to approximately geostationary range (22,236 miles). With new U.S. public and private sector operations extending into cislunar space, the reach of USSF's sphere of interest will extend to 272,000 miles and beyond—more than a tenfold increase in range." The "beyond" bit has infinite range.

This document clearly shows that while previously this was NASA's domain, it is now also that of the military. If the competition is going to be there, then Space Force will be there, but the area is huge. Keeping an eye on satellites in low Earth orbit is hard enough, but the major players will now also be trying to see what rivals are doing between there and the Moon.

The two are strategically linked. Total control of low Earth orbit by one nation could theoretically be used to prevent travel to cislunar space for others, and due to the vast distances involved, Earth-based radar and telescopes cannot monitor all the traffic that will be traveling between them. For now, they mostly track what is in low Earth orbit. They also can't see directly behind the Moon to the Lagrange point 2 region, where China's satellite gives it permanent oversight of the far side of the Moon, where it is considering building a base.

Positioning military satellites hundreds of thousands of miles up will give whoever does it first an advantage. They might be for monitoring, but competitors would be concerned that they're armed and could fire "down" at their satellites or even spaceships. There's not much point in building a base on the Moon if you can't get there and back because the opposition is in the way.

Space Force is ambitious. It says it will build the Cislunar Highway Patrol System. Its acronym "CHPS" probably harks back to a hugely popular and even more hugely cheesy 1970s TV traffic cop series, but fortunately for Space Force most people won't remember it. CHPS will involve a spaceship patrolling "far beyond the crowd" and will provide "critical national defense for the moon and beyond." These "celestial cops" could have a range of responsibilities. Carrying a large consignment of precious metals? We can escort you, ma'am. Driving dangerously? Pull over for a moment, sir. Out-of-control satellite traveling at great speed? Better put on the hazards.

Theoretically this would not extend to the Moon, because the Outer Space Treaty says: "The establishment of military bases, installations and fortifications, the testing of any type of weapons and the conduct of military maneuvers on celestial bodies shall be forbidden." However, it does permit "the use of military personnel for scientific research or for any other peaceful purposes" and allows the use of "any equipment or facility necessary for peaceful exploration of the Moon." It's not far from there to arguing that your military officers inside NASA, who are conducting scientific experiments on the Moon, require the means to defend themselves, what with the current situation being [insert current situation of any crisis year].

It's difficult to believe that the Big Three space nations do not have feasibility studies about building military bases on the Moon. After all, during the Cold War both the Soviets and the Americans looked at the possibilities. One declassified "secret" document on the American side described building an underground military base to house a Lunar-Based Earth Bombardment System. Something similar does not appear to be in the current strategies of the Big Three, but if any country begins occupying the strategic positions on the Moon where the water, helium, titanium, and other riches are, and then tells other countries to back off, a military stand-off would be likely. Tightly defined agreements and confidence-building measures are urgently required. Without them, the ideal of the Moon being for all of us will be reduced to moondust under the feet of a new generation of astronauts, cosmonauts, and taikonauts.

NASA—with Space Force behind it—is now on its way back to the Moon. There is crossover between the military and civilian US space activity, but both sides try to keep most of it separate. When it comes to astronauts, though, there's a limited pool of qualified candidates, and so traditionally most have come from the military and have been men. However, in 2020, the team of NASA astronauts for the Artemis mission to return to the Moon reflected the agency's efforts to diversify candidates' backgrounds. Of the eighteen people named, only ten are active-service military personnel, nine are women, and four are people of color. The intention is that the first woman and first nonwhite person to walk on the Moon will be American.

Skin color and sex are not the only differences from the last time humans were there. Another difference is computing power. When Armstrong made the first step, and Cernan the last, the computers used to get them there were several million times less powerful than your smartphone is today. But perhaps the biggest difference is that this time we are going there to stay.

The astronauts will get most of the way there in the Orion spacecraft on top of the Space Launch System—the most powerful rocket NASA has built. It's in competition with SpaceX's Starship, and although NASA would be reluctant to give up its giant baby, the rival is designed to be reusable and therefore cheaper. The plan is to construct the Gateway space station near the Moon and use it as a docking station for the Orion. The Gateway is a joint venture between NASA, the European Space Agency, and the Japanese and Canadian space agencies. Its modules will be delivered over several missions by SpaceX Falcon Heavy rockets. From Gateway, the astronauts can pick up the Human Landing System craft for the journey down to the Moon's surface. The return journey reverses the process.

Gateway is key to this plan. It will be situated on a highly elliptical orbit around the Moon. This means that at times it will be relatively close to its surface, facilitating landing missions, but at certain points in the orbit it will edge closer to Earth, making it easier to pick up astronauts and supplies coming from home. If the method works, it may be repeated for the plan to get humans to Mars. The idea is to decrease reliance on Earth.

Gateway will have a Habitation and Logistics Outpost (HALO) module where astronauts can live and conduct scientific experiments for up to ninety days between visits to the Moon. HALO will also be used as a communication relay system between Earth and the Moon, and to control rovers.

One of the most important experiments on board HALO will be measuring radiation levels. Once astronauts get beyond Earth's magnetic field, they are exposed to high-energy particles that can increase the risk of cancer and harm the central nervous system. The ISS is in low Earth orbit, which reduces the amount of radiation astronauts working there are subjected to. But Gateway will face much higher levels of radiation. It will be built to shield those living inside, but still obtain precise measurements about radiation over lengthy periods and their potential effects on the body.

By 2030, Gateway should be built, the test runs completed, and the first astronauts delivered to the Moon. The Artemis timetable has slipped a few times, but the successful launch of the uncrewed Artemis 1 mission in late 2022 meant that the heavy-lift SLS rocket passed its first test with flying colors. The Orion spacecraft it took up traveled 40,000 miles beyond the Moon, breaking the distance record for a spacecraft designed to carry humans. In the same year NASA's microwave-sized CAPSTONE spacecraft arrived in an elliptical orbit around the Moon to help determine where to build Gateway.

The lunar landing site has yet to be chosen, but it is expected to be near the south pole. This would be a first as the Apollo astronauts never went close to either pole. The scientists are still searching for the best place to situate the Artemis base camp, which at first would host the astronauts for a few days, but eventually would become a full lunar base with living quarters, radiation shields, communication systems, power infrastructure, vehicles, and a landing pad.

Given the amount of time the astronauts will need to spend on the surface, and the huge temperature swings between the sunlit and shaded areas, NASA has been working with private companies to design a new generation of space suits, rover vehicles, and cameras. The first American space suits were upgraded aircraft high-altitude flight suits. Each subsequent generation has built on the previous, and the latest iteration is a serious improvement on the ones currently used for spacewalks outside the ISS. NASA calls them the Exploration Extravehicular Mobility Unit, or xEMU for short, although Artemis Space Suits might have been easier.

At first glance they look similar to the Apollo suits we saw Buzz Aldrin and Neil Armstrong modeling, but the xEMUs are nowhere near as restrictive. They have significantly improved movement in the legs, waist, and arms and allow the wearer to actually walk on the Moon, as opposed to awkwardly bunny-hop, and to lift objects above the helmet. Previous suits absorbed exhaled carbon dioxide until they reached saturation point; the new version will absorb it and then pump it out into space. Miniaturization of electronics has allowed the backpack to house duplicates of key safety features, with warning sounds and lights in case of failures. The communication system in the helmet has been completely overhauled and includes an HD camera and voice-activated microphones connected to a high-speed data link . . . "NASA—play 'Homeward Bound' by Simon and Garfunkel."

The suits can withstand radiation and temperatures from +238°F to -248°F and are designed to provide full life support for six days in case of emergencies. NASA calls them "personalized spaceships." But despite all this twenty-first-century techno-wizardry, our intrepid explorers will still have to wear diapers.

The new rover vehicles are also designed not to leak. Called Space Exploration Vehicles (SEVs), they look nothing like their twentieth-century equivalent, the "Moon buggy." The new models will have pressurized cabins allowing two astronauts to drive long distances at 6 mph without wearing their suits, then put them on and get out for spacewalks.

This all costs. A lot. But relative to Cold War spending, it's cheap. In the 1960s, NASA's annual expenditure hit 4 percent of the federal budget; today it's roughly 0.5 percent. The difference is that beating the Soviets to the Moon was a price considered worth paying. Spending has also come down as NASA buys services from private companies that have innovated and cut the cost of rocket launches.

All stages of the Artemis program, from the launch rocket to the rovers, involve collaboration with private firms. Some companies are content with a supporting role in space exploration, but several intend to have their own missions and subsequent profit-making enterprises.

SpaceX has won the contract from NASA to build the lunar landing module that will deliver astronauts to the Moon from Gateway. It already ferries American astronauts to the ISS. In 2010, SpaceX became the first private company to launch, operate, and recover a spacecraft. Two years later it was the first private company to launch a craft that reached the ISS. In 2020, it launched Starlink, which, as we saw in chapter 4, delivers broadband signals and has become the largest satellite constellation. The following year it was the first company to put nonprofessional astronauts in space. When a SpaceX rocket goes up, its first stage comes back down about ten minutes later and, usually, lands, ready to be used again. The company has significantly lowered the cost of launching rockets and proved that start-ups can compete against heavyweights such as Boeing.

Elon Musk has plans. Big, big plans. As we've seen, they include putting astronauts on Mars—soon. Why go? According to Musk, "There's so many things that make people sad or depressed about the future, but I think becoming a spacefaring civilization is one of those things that makes you excited about the future."

Many people disagree. The eminent astrophysicist Martin Rees is not against spacecraft heading for Mars, but it's not a priority. He told the *Guardian* that Musk's ideas were a "dangerous delusion . . . dealing with climate change on Earth is a doddle compared to making Mars habitable."

Jeff Bezos also disagrees, and has different plans. The former CEO of Amazon and founder of Blue Origin wants to build cities, but closer to home. He argues that planets are not the best place for our expanding populations; instead, he aims to construct giant domed cities to orbit Earth. That's domed, not doomed.

In the shorter term, Blue Origin has designed a landing craft it hopes NASA will use once the Moon base has been established. The company is already in business ferrying tourists to space aboard its reusable New Shepard rocket, named after the first American in space, Alan Shepard. Bezos has taken the trip himself, as has William

Shatner, aka Captain James T. Kirk, who at ninety became the oldest person to reach such heights. On his return he was emotional to the point of tears, calling the journey his "most profound experience."

Blue Origin's massive New Glenn (after John Glenn) rocket is designed to carry up to 50 tons of cargo to low Earth orbit for paying customers, and Bezos clearly has plans beyond that. At one stage he was hinting at a New Armstrong rocket. Yes, Neil Armstrong.

Richard Branson's Virgin Galactic beat Blue Origin into space by more than a week, although Bezos didn't accept this. Branson's rocket, launched from a plane, took him up about 52 miles—roughly above what NASA defines as Earth's boundary. But the New Shepard went up 62 miles, above the Kármán line, the height accepted as space by the Fédération Aéronautique Internationale. Both companies are therefore right—according to where you draw the line.

Virgin Galactic is concentrating on suborbital tourism. At around $450,000 a flight the customer base is small, but perfectly rich. If Branson is correct, then there are enough multimillionaires out there for the company to turn a profit and move on to lowering prices for a mass market. That might seem optimistic, but it was just over a decade between the Wright brothers' first flight in 1903 and the first scheduled passenger airline service in 1914 (in Florida), and only another four decades until more Americans were traveling by air than by train.

Virgin Galactic and Blue Origin now have a rival for space tourism. Sierra Space is the new kid on the launchpad with its Dream Chaser spaceplane, which at first will be used as a supply craft for NASA, but eventually is supposed to give you the vacation of your dreams—or nightmares, depending on your disposition.

These companies underline how we are now well into the Commercial Space Age. Accessing space in vehicles built and owned by private corporations is a step change. Private enterprise is no longer just trying to make profits from satellite-related activity but is looking ahead to space tourism, long-distance transport services, mining the Moon and asteroids, and 3D print-manufacturing in zero gravity.

In 2010, Made In Space, Inc. (MIS), was a two-room start-up business in California. Four years later a MIS Zero-G printer was flown to the ISS, where astronaut Barry "Butch" Wilmore unpacked it and printed the first part ever manufactured in space. Okay, it was only a faceplate for the printer itself, but it was quite a first. Later, Wilmore realized he needed a specific ratchet wrench. Back on Earth MIS typed a few lines of code and transmitted them to the ISS, where Wilmore duly printed out the wrench. MIS now has a $74 million contract with NASA to 3D-print large metal beams in space. It's a lot cheaper than flying them up there.

MIS is one of over five thousand American space-related companies. They tend to be more innovative than state entities and prepared to take risks. Private enterprise has succeeded in dramatically cutting the costs of space travel, which in turn has helped NASA to aim high.

NASA has always cooperated with commercial companies, but the plethora of start-ups and their ambitions has taken this to another level. It has signed deals with several whereby it will pay them to collect lunar soil. The fees are negligible—one company only asked for $1 to win the contract—but the deals benefit both sides. The companies get to practice extracting resources and NASA creates what it will then argue are business and legal norms for operating commercially on the Moon.

In late 2022, the Japanese company ispace launched its Moon lander on top of a SpaceX rocket en route to the lunar south pole to search for water ice. NASA has signed the rights to "own" whatever is found, which again raises the question of who owns the Moon. Japan, the UAE, and Luxembourg have passed legislation allowing their companies to be involved in such transactions, and the Americans passed a similar law in 2015 under the Obama administration. Private businesses so far have no advanced plans for their own Moon bases, but presumably US, Chinese, and Russian firms would take advantage of any "sovereign" bases built by their own countries.

NASA is working on a range of smaller projects, such as solar sail propulsion for robotic deep space exploration and laser communication systems, but the focus is on Artemis, Gateway, and a Moon base.

It will probably be built before the argument about "why" the US should go back is resolved. But as things stand, the realities of geopolitics, and now astropolitics, point to both the US and China going to the next stage in great power rivalry. If either does not, then the way is open for the other to "own" the Moon. The lunar water and rare-earth metals to be found there are not renewable resources.

It has been a long time since Americans were last on the Moon. The six American flags planted on the surface are now bleached white by the rays of the Sun. NASA's Lunar Reconnaissance Orbiter spotted five of them, still standing, in 2012. The Apollo 11 flag was knocked over when Aldrin and Armstrong took off. The flags are made of nylon and are likely to disintegrate within a few more decades. We should retrieve the Apollo 11 flag and put it in a museum and, while we're at it, find Armstrong's footprint and preserve it as the ultimate "Walk of Fame." Now there's a reason to go back.

Did I mention Russia?

The Russian Soyuz TMA-14M spacecraft after landing near the town of Zhezkazgan, Kazakhstan, on March 12, 2015.

7

RUSSIA IN RETROGRADE

Earth is the cradle of humanity, but one cannot stay in the cradle forever.
—Konstantin Tsiolkovsky, "Father of Cosmonautics"

Russia has shown it is able and willing to fire rockets at densely populated civilian areas, but its world-renowned prowess in firing them into space may be faltering. The two things are connected.

In February 2022, on the same day Russian forces invaded Ukraine, the American government announced wide-ranging sanctions against Moscow. Among them were those intended to "degrade their aerospace industry, including their space program," with embargoes on semiconductors, lasers, sensors, and navigation equipment.

Dmitry Rogozin, then chief of the Federal Space Agency, Roscosmos, was unimpressed. Russia and the United States have collaborated on the ISS since 1998, but in a tweet to his 800,000 followers he said, "If you block cooperation with us, who will save the ISS from an uncontrolled deorbit and falling on US or European territory?" The Russians control the propulsion required to prevent the station from falling back to Earth, while the Americans supply the life support system.

It was standard fare. Previously, Rogozin had demonstrated his nationalist credentials by suggesting that American astronauts should try getting to the ISS on trampolines rather than Russian rockets, which they had been using for some years. The day after the sanctions were imposed he switched vehicles, saying: "Let them fly on something else: their broomsticks."

America's SpaceX hit back. Elon Musk's company was already working to deliver its Starlink satellite internet service to Ukraine, as we've seen. On March 9, a SpaceX Falcon 9 rocket carrying a batch of satellites was seconds away from liftoff. Viewers of the event's live feed heard launch director Julia Black tell her team, "Time to let the American broomstick fly and hear the sounds of freedom."

Mr. Rogozin called veteran US astronaut Scott Kelly a moron, hinted that Russia might leave a NASA astronaut behind on the ISS, and published footage of technicians taping over the American flag on a Soyuz rocket. Back came Kelly: "Without those flags and the foreign exchange they bring in, your space program won't be worth a damn. Maybe you can find a job at McDonald's if McDonald's still exists in Russia." It doesn't.

At one level it was all good knockabout stuff, but at another we were watching a decades-old space partnership crash and burn, ending a relationship that had been beneficial for science, détente, and humanity. The geopolitical fault lines in space were being redrawn. The events of 2022 make it more likely that Russia will step away from exploration and concentrate on military applications in space. They also accelerated the division of space activity into two blocs: one led by China, the other by the US.

The repercussions were widespread. In the immediate aftermath of Russia's invasion of Ukraine, and subsequent sanctions, Moscow said it would no longer sell rocket engines to the US, but this was a limited blow to the Americans given that they were already weaning themselves off relying on Russia for most things space-related. It announced it would cease working with Germany on joint scientific experiments aboard the ISS. The Germans stopped all scientific cooperation with Russia, which included switching off a German-built space telescope that was hunting for black holes in a joint Russian-German project.

Roscosmos halted launches of Soyuz rockets from Europe's spaceport in French Guiana and withdrew its workforce. The Kourou spaceport is where high-profile missions such as the James Webb Space Telescope had been launched. The suspension delayed the European Space Agency's (ESA) ExoMars program, which had been due to launch a mission to the Red Planet. The ESA officially ended its relationship with Roscosmos on July 12 and began to look for a new way to get Mars-bound. The last straw may have come a few days before, when Roscosmos published photos of cosmonauts on board the ISS holding flags of two regions of Ukraine occupied by Russian forces.

Roscosmos also announced that it would not launch thirty-six satellites for the London-based OneWeb unless it guaranteed they would not be used for military purposes. They had been due to launch from the Russian-run Baikonur Cosmodrome in Kazakhstan. It also demanded the "withdrawal of the British government from the shareholders of OneWeb," as the UK had helped the company avoid bankruptcy in 2020. OneWeb refused and said it was suspending all launches from the site. SpaceX then helped OneWeb launch its satellites despite being a competitor.

There were many losers from this sorry episode of tit for tat, including Dmitry Rogozin, who was dismissed from Roscosmos a few days after the ESA cut ties with it. But the biggest loser was Russia and its space program, which looks set to decline.

Russia had already been losing market share in the competition for sales of rocket engines, satellite launch services, and delivery of astronauts to the ISS. After the US Space Shuttle fleet was retired in 2011, astronauts had to hitch rides on Soyuz craft to get there (hence Rogozin's jibe about "broomsticks"). But since 2020, NASA has also had the option of SpaceX Dragons with which to make the journey.

Because of the unique situation on board the ISS, a working relationship had to be maintained, even amid the furor, but now Russia does not appear interested in helping NASA extend the station's working life up to 2030. Given the dire state of relations between Moscow and Washington, it is highly unlikely that NASA will invite Roscosmos to collaborate on the US-led Gateway, nor will private American companies be rushing to partner with it on the multiple commercial space stations that are in concept development.

Russia has been cut off from most of the planet's space cooperation, funding, and expertise just as the sector is expanding more rapidly than it has for decades. Russia's best days in cosmology look to be behind it; its future may be as the junior member in a Sino-Russian partnership.

Following Moscow's invasion of Ukraine in February 2022, more than a million Russians fled their country, among them thousands of engineers, computer experts, and scientists. It's not known how hard this has been on the Russian space program, but it will surely affect it. Several unconfirmed reports in the weeks after the war broke out suggested that Roscosmos had banned its employees from going abroad in case they did not return and that border guards were under instructions to stop certain categories of scientists from leaving the country. The American and British governments then proposed legislation making it easier for Russian scientists, including those in the space industry, to get work visas and permanent resident status. It is a far cry from when the red star shone so very, very brightly in the firmament of science and human endeavor.

———

The Soviets achieved a number of stunning firsts, from Sputnik to the first man in space, and even after they lost the race to get to the Moon their spacefaring prowess continued. On and out they went, as far as Venus and Mars, and up into low Earth orbit with a series of space stations, including the very first, Salyut 1, in 1971, as they concentrated on technologies to allow a long-term human presence in space. But their success was not to last.

The Soviet Union was dissolved in late 1991 and early the following year the Soviet space program was replaced by the Federal Space Agency, which eventually became Roscosmos. With the economy in turmoil, the government made severe cuts to the space budget throughout the 1990s, despite its leading role on the ISS.

And that hasn't all been sweetness and bright light. A recent series of incidents involving the ISS has outraged Russia's partners.

In 2018, the Russian state-owned TASS news agency published an extraordinary article about the US astronaut Serena Auñón-Chancellor. Without offering any proof, it effectively accused her of having "an acute psychological crisis" on board the ISS and drilling a hole in a docked Soyuz capsule. The reason? According to the defamatory TASS report, it was because the hole would have slowly depressurized the whole station, requiring her to be flown back home immediately.

There was indeed a hole, and it was patched up. Where and when it happened has not been established, leaving open the possibility that it had occurred on the ground. But the idea that an American astronaut, in space, had caused it deliberately was beyond ludicrous and suggested that someone somewhere was trying to shift the blame. The Russians even sent two crew members out on a spacewalk to gather "evidence"; cosmonaut versions of detectives Clouseau and Poirot took knives with them and cut away some of the insulation from the outside of the Soyuz craft to investigate the "scene of the crime." The official Russian report into the incident has not been released.

In 2021, there was an even more dangerous occurrence. The good news: Russia's twenty-two-ton Nauka laboratory module successfully docked with the ISS. *Nauka* means "science" in Russian, and the module gave Roscosmos a plethora of new experimental capabilities. And an extra toilet. The bad news: Three hours after docking Nauka's thrusters began firing, sending the whole station into a cartwheel. American and Russian mission control liaised and began firing thrusters on the other side of the station to bring it back under control. The emergency went on for almost an hour and ended only when Nauka ran out of fuel. Roscosmos said little about the incident, but eventually blamed it on Ukrainian-built machinery in Nauka's propellent tanks.

Neither of these events, however, matched the 2021 incident when Russia destroyed one of its obsolete satellites, sending space debris hurtling toward the ISS. The international space community lined up to condemn the action. All these events coincided with the deterioration in cooperation between Russia, the US, and European powers. The arc of the relationship was already heading downward even before Russia's seizure of Crimea in 2014, but the annexation of what legally remains Ukrainian territory accelerated the trajectory.

President Putin has made no secret of his desire to reverse the effects of the collapse of the Soviet Union, an entity he's described as "another name for Russia." With all the former Warsaw Pact coun-tries joining NATO at the first opportunity, he watched with alarm as, in his view, NATO advanced toward Russia's borders.

This century he has worked to restore Russia as a world power, mostly through its military. After the Soviet Union's forces were dis-banded, Moscow created the Russian Space Forces in 1992. This went through several iterations and is now a sub-branch of the Russian Aero-space Forces. Bringing the two together was part of an attempt to create an efficient single command with responsibility for all military aspects of aerospace. In this aspect it was four years ahead of the US. According to its web page, the Russian Space Forces are tasked with monitoring space for incoming threats, prevention of attacks, building and launch-ing spacecraft, and controlling military and civilian satellite systems.

In 2003, the senior command of the Russian Aerospace Defence Forces had watched keenly as the US military sliced through Iraq's half-a-million-strong army using satellites to precisely target troops, equipment, and buildings. By the time the US ground forces rolled in, Iraq's army was in no shape to resist.

Analysts noted that during the Second World War, 4,500 air sorties had been required to drop 9,000 bombs to destroy a railway bridge. In Vietnam the figure was 190 bombs; in Kosovo it required only one to three cruise missiles. By the time of the invasion of Iraq, a single missile guided by satellite could do the job. Moscow realized it had fallen behind the US's space-based military assets and set about trying to catch up.

It currently operates the GLONASS global positioning system, which is the Russian equivalent of America's GPS.

The GLONASS constellation of twenty-four satellites was fully completed in 1995, a year after GPS. To remain at full capacity to have global coverage requires frequent launches of new satellites to replace those that break down or reach the end of their shelf life. However, in the economic chaos of 1990s Russia, funding for space projects was cut by 80 percent. By the end of 2001, only six satellites were operating, not enough to even cover Russia. This was a serious blow to Moscow's strategic interests as GLONASS ensures that its nuclear missiles can find their targets.

After Putin came to power in 2000 the economy began improving, and he made the restoration of the system a top priority, more than doubling its budget. By 2011, it was back to twenty-four satellites, giving global coverage for the first time in a decade. Sanctions complicate Russia's ability to persuade phone and car manufacturers to enable GLONASS in their products but its military capabilities remain intact and the accuracy of the system is not in question.

The focus on GLONASS demonstrated the military's concern that it needed the situational awareness and communication reliability that only a satellite-based system can provide. GLONASS has been used to support Russia's military operations in both Syria and Ukraine, which have used high-precision weapons. This resulted in Ukrainian hackers targeting GLONASS, but with limited success. As the Kremlin became reliant on these systems it was logical that it would invest in defending them.

It has also invested in its ability to attack its enemies' satellite systems. A way to do this is by using one of your satellites to get close to someone else's. There are many legitimate reasons for doing so: for example, to inspect damage caused by debris. But you might also want to grab hold of it, blind it with a liquid substance, or even fire at it. On several occasions the US has made official complaints that Russian satellites were "stalking" American ones. In 2020, U.S. Space Command was concerned to see Russia's Cosmos 2542 satellite release another satellite from inside it. Instead of staying close to other Russian satellites, 2543 approached an American military reconnaissance vehicle. More alarming still, it went on to fire a high-velocity projectile into outer space.

As this event indicates, Russia is building a range of options to give it warfighting capabilities in space. Some of them are dual-use facilities that allow plausible deniability of military intent; others are justified as being deterrents to prevent war.

As well as using satellites as weapons, Russia and other countries are working on land-based weapons to fire into space. The 2021 ASAT test firing is one of a string of examples showing that Russia, aware that it cannot match the US militarily in space, seeks to demonstrate its ability to disable or destroy its adversary's core equipment. The obsolete satellite it blew up was one of its largest. There were many others it could have selected that would have produced far less debris. It chose to send a message. From the Kremlin's perspective this is a rational insurance policy.

The same applies to a project to launch a rocket into space from under a modified MiG-31 fighter jet already flying at supersonic speed. The rocket is then thought to be able to release a small satellite, possibly one capable of firing a weapon.

One weapon already operational is the Peresvet laser system, designed to counter satellites. These are truck-mounted devices deployed with five of Russia's mobile intercontinental ballistic missile divisions in order to target foreign satellites as they pass over Russian territory and prevent them from tracking the units' movements. It's unclear if they can "dazzle" or "blind." "Dazzling" means swamping a satellite with so much light it temporarily loses sight of what it's trying to see. "Blinding" permanently damages a satellite's imaging system. It's not known if any of the five units have used them successfully.

Most analysts believe that Peresvet can only dazzle, but the authoritative online magazine *Space Review* suggests Russia is ready to step it up a notch with a new system known as Kalina. An in-depth investigation in 2022 looked at Google Earth images and open-source patent documents and found that Russia's Krona space surveillance complex was working on a state-of-the-art laser system with the ability to destroy a satellite.

The Krona complex, atop a 1.2-mile-high hill, is just west of Zelenchukskaya, near the border with Georgia. New ground has been broken and a dome designed to house a telescope has appeared. According to *Space Review*, technical documents for the tender to construct it describe a building able "to operate in temperatures ranging from +40 to -40°C and withstand magnitude-7 earthquakes." The dome consists of two sections that can be opened in less than ten minutes, allowing the telescope to scan the entire sky.

The building is connected via a tunnel to another that houses a LiDAR (Light Detection and Ranging) machine. The LiDAR pulses light toward a satellite and then measures how long it takes for each pulse to return. This gives it an indication of the satellite's position and the direction and speed in which it's traveling. The more sophisticated the equipment, the more accurate the reading.

If Kalina is operational, at this point it would begin to focus and fire. The laser beam must pass through Earth's atmosphere and so needs to be powerful. The more light it delivers, the more damage it can do. Most observation satellites operate just a few hundred miles up, in low Earth orbit. It's thought Kalina will be able to lock onto and track a satellite for minutes at a time, and either dazzle or blind it during these periods. *Space Review* estimates that the system could enable Russia to shield from view about 38,610 square miles of its territory at any one time—an area bigger than Portugal.

Kalina would also be able to select a spot on the satellite and focus all the laser's energy on it. This could burn out the machine's cameras or its engines, rendering it useless. Lasers with this much energy are thousands of times more powerful than those used to play CDs or in surgery, and Kalina can ensure that multiple beams fired from a telescope with a diameter of several feet travel in parallel with each other and don't spread out. If it works, Kalina can probably take out satellites as high up as geostationary orbit.

If deployed, their use can be denied. The laser beam is invisible, there's no loud bang as it's fired, and no plume of smoke afterward. "What's that?" says Moscow. "Lasers? Act of war? Nothing to do with us. Have you tried North Korea?"

Now imagine such weapons being fired from space. Not into space, but from space. Without any atmosphere to deflect or weaken the beam, the weapon could be much smaller and the target bigger— a space station, for example.

Kalina is among the new generation of systems that has been dubbed Putin's *superoruzhie*, or "superweapons." They include hypersonic missiles that have the ability to change direction and altitude while traveling in Earth's atmosphere. This would make it difficult for the target country to know where a missile is headed and prepare accordingly.

Since 2018, Russia's military efforts in space have been closely linked with China's in a bid to undermine US space superiority and threaten its infrastructure. The relationship began in the early 1990s. Sanctions on various technologies had been imposed on China following the 1989 massacre of pro-democracy protesters in Tiananmen Square, Russia had emerged from the wreckage of the Soviet Union, and so Beijing and Moscow slowly began to cooperate on space policy.

By 2018, they were ready for a formal agreement to cooperate on a range of projects including rocket engines, spaceplanes, satellite navigation, and monitoring space debris. (But, as we've seen, the last is not necessarily as benign as it sounds because if you have a monitoring system you also potentially have a spying system.)

This, and the advancement of space-related weapons, is why the Americans and the Europeans have for years been suspicious of joint Russian-Chinese proposals for a new treaty to prevent a space arms race. The texts of drafts put forward in 2008 and 2014 are still being discussed and are notable for what is missing.

The texts may be peppered with references to "peaceful purposes" and "arms control," but, like all other proposals and agreements so far, they do not define what constitutes a weapon in outer space, nor detail any limitations on how close one nation's satellites can be to another's. More seriously for the Americans is the lack of clarity on the developing, testing, or stockpiling of ground-based anti-satellite weapons such as Kalina. This suits Moscow and Beijing. They know they are behind the US in conventional warfare ability, and that modern conventional warfare relies on satellites. Therefore, they are not interested in banning the weapons that could shoot at those satellites from Earth.

As we've seen, the United States has proposed a worldwide ban on direct-ascent anti-satellite weapons, which can cause space debris, and has called for a more encompassing treaty to address the new issues brought by new technology. But it's difficult to see how a consensus is going to be reached, especially as the Americans are developing their own ground-launched weapons and other technologies even if they have announced a unilateral moratorium on direct-ascent missiles.

A more likely scenario is that Russia and China will continue to develop their relationship with initiatives such as their plan to build an International Lunar Research Station "on the surface and/or in the orbit of the Moon" by 2035.

As part of various transfers of technological know-how, the two countries have been working to make Russia's GLONASS and China's BeiDou satellite navigation systems compatible. This means that if one country went to war with a third party and its communication and observation system was damaged, it could use the services of the other.

It sounds like a win-win, but . . . Putin . . . we have a problem.

Russia is the junior partner in this relationship and Russia doesn't want to be the junior partner in anything. Moscow has the history, the legends, and the medals to show it. But Beijing has the money and the infrastructure, and it's not playing catch-up anymore. The old cliché that Chinese space technology is reengineered Russian space technology is well out of date. China is the one with its own space station now, not Russia. China has landed a craft on the far side of the Moon, not Russia. It's also ahead on technology for heavy-lift reusable rockets, and its space-related private sector is more vibrant.

Russia needs China more than the other way around, which means that Beijing can afford to be cautious when it comes to helping Moscow. The Chinese are reluctant to supply Russia with technology due to economic sanctions—if doing so would trigger sanctions against China as well.

Despite Chinese hesitancy in their "friendship," the relationship is useful to Russia. After pulling out of the ISS, the only place to have cosmonauts in space for a long period of time will be on board the Chinese space station. Without China, Russia would not be able to afford to build its own base on the Moon. The partnership allows Russia to try to compete as a major space power and allows China to buy oil and gas at "partnership" prices. Behind the deal lies the joint strategy of building an alternative power bloc to the American-led loose coalition of democracies, and then persuading other countries to join them. But when it comes to Russia's space program—that's an offer most can refuse.

———

Russia was once cutting edge; now it's being cut out. There's also an element of cutting itself off. New laws mean that any Russian media outlet that reports even basic information about the country's space industry must add a disclaimer to the article/tweet/post reading: "This Report (Material) has been created or distributed by Foreign Mass Media executing the functions of a Foreign Agent, and/or a Russian legal entity executing the functions of a Foreign Agent." Declaring yourself a "foreign agent" has never been a good idea in Russia at the best of times, and these are not the best of times.

The Russian public, which still has a keen interest in space, will be denied information on almost everything about it except the most asinine government-approved particulars. A 2019 opinion poll found that 31 percent of Russians closely follow news about space.

Some 59 percent wanted the country to maintain its stellar efforts, and 53 percent believe it will.

And it's clear that, despite its decline, Moscow is making plans to stay in the top league.

Russia's new crown jewel is its most modern space launch facility: the Vostochny Cosmodrome. In 1991, post-Soviet Russia did not have a major spaceport on its territory and had to pay Kazakhstan for the privilege of launching from Baikonur. It was determined to remedy this embarrassing situation and the Kremlin has bet on Vostochny as the answer. It intends to develop strategic autonomy by breaking with the Soviet past and ensuring that all the major components of its military and civilian space projects are based within its borders.

Construction began in 2007 in Amur Oblast in Russia's Far East. It's about 5,000 miles from Moscow and 124 miles from the border with China. The nearest city is Blagoveshchensk (population 200,000), which sits on the northern side of the River Amur. It's a typically drab former Soviet municipality from where inhabitants can see the shiny new Chinese city of Heihe, with its neon lights blazing across the river from modern high-rise apartments and office blocks. Fifty years ago, Heihe was a sleepy village; now it is home to 224,000 people and stands as a reminder of how China has overtaken Russia.

That's where Vostochny comes in. The cosmodrome project is intended to have an economic knock-on effect across Amur Oblast, which is one of Russia's most underdeveloped and isolated regions. It was chosen for both economic and geographical reasons. It's on the site of a former intercontinental ballistic missile base and so has access to existing main railway lines. Its remote location reduces the risk of rocket debris hitting any large urban center, and the latitude means rockets can carry almost the weight of those launched from Baikonur.

It is situated near part of the Trans-Siberian Highway, which supports the infrastructure required for such a massive project, and involves the creation of a new city for 35,000 people.

Vostochny, completed overbudget and behind schedule, was plagued by the endemic corruption that dogs all Russian industry. President Putin, who takes a keen interest in the misappropriation of state funds, has reminded senior politicians that Vostochny is "practically a national project. But no, they keep stealing in hundreds of millions!" At least $172 million was looted by top officials, dozens of whom were arrested and jailed.

Rockets are now being launched from the site, but completion of several smaller projects may take at least another decade, so there's plenty of time to steal more money. A plaque at the site's main entrance declares: "The path to the stars begins here." Well, no one would dare have added "(unless the money runs out)."

The ambition is there, but the finance, the equipment, and possibly the expertise required to match the American and Chinese space programs may not be. Despite this, several other long-term projects are currently under way.

A reusable two-stage rocket is planned to launch from Vostochny by 2026. Named Amur, it looks suspiciously similar to SpaceX's Falcon 9 but is smaller and the reusable section will be able to carry only 10 tons of cargo. That's an upgrade on Soyuz 2 rockets, but still less than half what Falcon 9s can take up.

The designs for a new space station named the Russian Orbital Service Station (ROSS) are complete, but the target for having it in orbit has slipped from 2025 to 2028, and some Russian experts are talking about 2030. Given that it took a decade for Russia to design, build, launch, and attach its Nauka laboratory module to the

ISS, even that seems optimistic. Nauka was supposed to be in service in 2007 but first docked in 2021. If it is built, ROSS will be smaller than the ISS and only inhabited for four months a year, limiting the amount of research cosmonauts can do.

There are also plans to build a "space tug" that will get an uncrewed spaceship to Jupiter (via the Moon and Venus) in just over four years. It will feature laser weapons and a 500-kilowatt nuclear reactor to power its electric engines. Named Zeus, its first mission is to launch in 2030. A mock-up of the craft displayed at the 2021 Moscow air show looked like a giant Erector set and had similar flying potential, but if the technology does get off the ground, then four years to Jupiter is feasible, as would be a two-year crewed roundtrip to Mars.

A space station, a reusable rocket, a space tug—it's an impressive list. Now all they need to do is find the funds, scientists, and equipment to take it off the list and into space.

Even before the invasion of Ukraine, Russia was losing revenue from its space activities—as we've seen, it was facing increasing competition for its cosmic taxi service. Given that it was charging $70 million per passenger to get foreign astronauts to the ISS, that severely dented the revenue stream. The US is also phasing out buying Russian-produced rocket engines and purchasing made-in-America versions.

Russia doesn't publish the budget for its military space program, but open-source reporting suggests it is in the region of $1.5 billion a year. Roscosmos funding has been cut to about $3 billion a year, with almost nothing earmarked for research and development. By comparison, NASA's James Webb Space Telescope project alone cost $10 billion, NASA's annual budget runs at about $25 billion, and US government annual spending on military space activities hit

$26.3 billion in 2023. China spends far less, in the region of $10 billion, but appears to be committed to increasing that.

What's more, Russia's space program is plagued with systemic problems, riddled with corruption and, aside from the Vostochny Cosmodrome, relies on an obsolete infrastructure, some of which is beyond the country's borders. Domestic private enterprise is reluctant to invest in a high-risk, state-dominated industry, which it knows is being squeezed by sanctions.

Add to this an aging population. A large proportion of the experienced Russian workforce is approaching retirement and so the industry will require at least 100,000 highly trained specialists to replace them this decade. However, talented young Russian engineers and scientists are not attracted to an industry that pays less than other high-tech ventures.

With increased sanctions hitting the Russian economy, making it difficult to source materials, Roscosmos will struggle to compete. Russia will not rest, nor accept secondary space-power status, but without the means to stay in the top tier of space exploration and scientific research it will settle for top-tier military applications.

The necessity of cooperation is what kept the air locks open between the Russian and American space communities even when the state-to-state relationship was broken. Routes to détente are not always easy to find, but there's one shining so brightly that we can see it with the naked eye as it passes overhead every ninety minutes traveling at 4.7 miles a second—the ISS.

However, the geography of space is not immune to the geopolitics of Earth. The détente of the Soyuz–Apollo docking, and of the ISS, is now lost in the space between us.

8

FELLOW TRAVELERS

There are no passengers on Spaceship Earth. We are all crew.
—Marshall McLuhan, philosopher

While China, the US, and Russia are the three main players in space, many others are looking to increase their presence. New technologies have delivered easier access to opportunities for an increasing number of countries, including those in the developing world. But due to the costs and infrastructure requirements, most cannot launch independently. This is part of a move toward the creation of space blocs.

The Europeans have a head start. The European Space Agency was formed in 1975 by ten nations and now has twenty-two members. It's dominated by EU countries as part of the aspiration for "ever-closer union," but is a separate institution and sits outside the EU's space program. The EU contributes about 25 percent of the ESA's budget, but individual member states pay for most of the rest. To their credit, the member states have captured about 20 percent of the world market in commercial space enterprises, such as satellite-building and

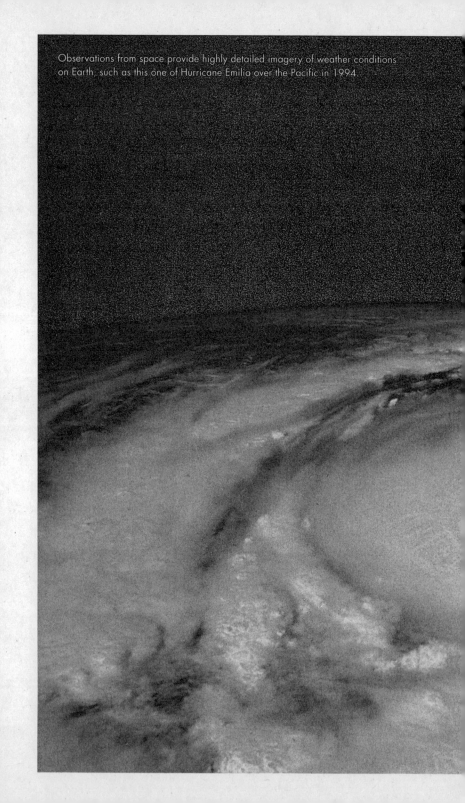

Observations from space provide highly detailed imagery of weather conditions on Earth, such as this one of Hurricane Emilia over the Pacific in 1994.

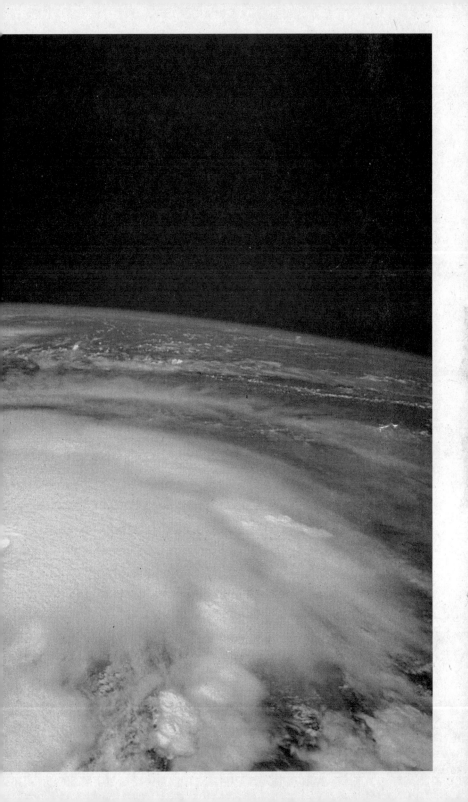

launches, robotic arms, and habitation modules, which is quite an achievement given that its budgets and private investment levels are almost loose change compared with American spending.

As an organization, the ESA has had some signature successes, including the Galileo global navigation satellite system; Copernicus, its Earth observation program; and its role in the ISS. However, for a project as ambitious as building a Moon base, the Europeans need to cooperate not only with each other but with a major power—in this case the United States. The ESA is firmly allied with the US and is part of the Artemis Accords Moonshot.

So while the Russians sent the first dog into space, the Europeans can claim the first sheep. Shaun the Sheep to be precise. Okay, not a living sheep, but nevertheless a fitting representative of planet Earth given that he bleats in most of the world's major languages—"Baa" in English, "Bee" in French, "Meeh" in Japanese—and has been seen in 180 countries.

Stuffed toy Shaun was on board the Artemis 1 flight that launched from Florida's Kennedy Space Center in November 2022 and flew 43,500 miles beyond the Moon before returning to Earth. It was the first integrated test of the Orion spacecraft and NASA's heavy-lift Space Launch System rocket. Shaun was chosen for the mission by the ESA, which built Orion's life support system—providing power, water, and oxygen, and keeping the craft on course. As the ESA's David Parker said, "Although it might be a small step for a human, it's a giant leap for lambkind."

As the mission progresses, we can expect fierce competition from the non-American Artemis signatories to ensure that one of their astronauts books a ticket, but as the Europeans are putting in most of the non-US money and have played a vital role in the project, the ESA believes they are "guaranteeing seats for ESA astronauts to explore our solar system." It didn't say which ones. To be the first European on the Moon will be a big deal; pan-European solidarity has its limits, and each country will want its astronaut to take that "giant leap."

Looking ahead, there is an ambitious plan for a 2029 launch of a spacecraft called Comet Interceptor, whose job is to do what it says on the tin. The idea is to park the craft near the James Webb Space Telescope at Sun-Earth Lagrange point 2, about 932,000 miles from Earth. It will hang about, waiting for a comet to approach that is new to our solar system and then intercept it. A comet making its first-ever orbit around the Sun would be gold dust because, as Michael Küppers of the ESA explains, it "would contain unprocessed material from the dawn of the solar system" that would help us understand "how the solar system formed and evolved over time."

However, despite examples of their undisputed expertise and world-class equipment, the Europeans are beginning to lag behind when it comes to space security. The first meeting of EU defense ministers took place in 2022; 2012 would have been late, but 2022 is careless. The EU as an institution is as dependent on space-based assets as any other major economic player, but it lacks the means to defend them. It keeps talking about the "need to guarantee our ability to operate securely" in space, but there's rarely progress on actually building anti-satellite weapons, directed-energy guns, or jammers. The European Commission (EC) is doing work on tracking space debris and on ultra-secure quantum-encrypted communication devices, but, again, rarely says how it intends to defend them. When the French discovered a Russian satellite approaching a Franco-Italian military satellite in 2017, their first thought was not "What will Brussels do about this?" Defense Minister Florence Parly said, "It got close. A bit too close." She accused Moscow of trying to intercept the extremely high-frequency-band communications used by French and Italian forces around the world. She also said that France had taken "appropriate measures," though did not go into further detail. An EU space force is about as likely as an EU army; more probable is that individual states will build their own capacity.

Three EU members have already forged ahead with their own space commands and policies—Italy, Germany, and France, which is the leading European space power. They all participate in the EU's space program but are not waiting for endless EC communiqués suggesting that "working parties might be formed to look at the possibility of holding high-level meetings to examine the feasibility of holding a summit to examine if the member states would back an EU spaceplane to be launched at T counting the twelfth of Never." Instead, all three are actively looking at building their own.

As in other areas, the Europeans want strategic autonomy: a policy that goes back decades at the nation-state and EU levels. In the 1960s, the democratic countries in mainland Europe could either rely on an American monopoly in space to provide for their needs or they could seek to build a level of domestic capacity.

The Italians were first up. In 1964, they launched the San Marco 1 satellite, albeit on top of an American rocket, and by doing so became the fifth country to have a machine in orbit after the USSR, the US, Canada, and the UK. Italy has since played a significant role in the development of the ISS and contributed several astronauts, including Samantha Cristoforetti, who was the first European woman to command it. Italy's giant defense company Leonardo has partnered with the French company Thales Group to form Thales Alenia Space, the largest satellite manufacturer in continental Europe. Thales is now working with Space Cargo Unlimited, based in Luxembourg, to build the first floating space factory, with the ambition to manufacture items related to biotechnology, pharmaceuticals, agriculture, and new materials. And in an effort to maintain the country's prowess the Italian government funded a 2021–27 budget worth almost €5 billion, including €90 million aimed at helping start-ups.

Next up was France, the fourth country to design, build, and launch its own rocket. After the Second World War President Charles de Gaulle had refused Washington's offer for France to shelter under the US nuclear umbrella. The idea of having American missiles on French territory was anathema to a man determined to rebuild French power. In 1964, French nuclear weapons became operational, and the following year a military communications satellite lifted off in a massive display of Gallic independence from *les Américains*, so now de Gaulle's "Force de Frappe" (strike force) had a ballistic missile that could deliver nukes and satellites. The A-1 (Army-1) satellite was quickly dubbed Astérix, after the popular comic-strip character Astérix the Gaul, who so successfully resists foreign domination.

The reflex of French exceptionalism was seen again in the 1980s. Colonel Muammar Gaddafi's Libya had invaded northern Chad in 1978. In 1983, Washington pressured Paris to intervene on Chad's side and provided the French military with satellite imagery of a quality France did not possess. The French were suspicious that some of the pictures were old and being used to draw them in, and the disagreements between the two powers led Paris to conclude that it did not want to rely on American reconnaissance. At the time French jets had to fly ten-hour missions to get images the Americans could capture in seconds, and so in 1986, France launched SPOT 1—a commercial imaging satellite capable of taking quality color photographs with sixty-five-foot accuracy. SPOT 1 was launched just in time for it to capture images of the Chernobyl nuclear accident, giving France a better view of what was unfolding than its European neighbors.

The year 1995 saw the launch of the Hélios military satellite, developed with Spain and Italy, and capable of three-foot resolution for black-and-white images. It was to prove useful in 2003 ahead of the Iraq War, when it provided intelligence that helped to persuade France not to participate in the invasion of Iraq.

Hélios has now been replaced by the dual-use Pléiades system, operated by Airbus, which provides information to commercial customers and the French and Italian ministries of defense, who each have a daily image quota reserved. This came in handy when France intervened in Mali in 2013 to stop an advance by rebel forces on the capital, Bamako.

France remains a major world space player both commercially and militarily. In 2019, it published its military space strategy, which declared: "France is not embarking on a space arms race." This was somewhat contradicted by later proposals for swarms of nanosatellites to guard bigger satellites, a ground-based laser system to blind opponents, and, somewhat implausibly, machine guns on board a satellite. However, Paris has ruled out building its own direct-ascent anti-satellite missiles on the grounds that it is irresponsible to add to the amount of space debris in low Earth orbit. Its space command was set up in 2019 near Toulouse—home to companies such as Airbus and Thales. Its brief is to guard French satellites and deter aggression against the country's space capabilities.

Unilateralism has its limits, especially as we slowly move toward another form of bipolar world, this time dominated by the US and

China (with Russia as junior partner). France's "go it alone" policy has therefore adapted to the twenty-first century. For example, France has joined the Combined Space Operations initiative, which initially was part of the Five Eyes intelligence-gathering alliance of the US, UK, Australia, New Zealand, and Canada. It is also increasing cooperation at a commercial level with the ESA.

Despite Germany being the first country to launch a rocket into space (Wernher von Braun's V-2), its position as a player in the industry flies under the radar. However, its space sector is the second biggest in Europe and it is the second-largest contributor to the ESA. The ESA's European Space Operations Centre is based in Darmstadt, near Frankfurt, from where its uncrewed spacecraft are controlled and debris is tracked. Munich is home to the Columbus Control Centre, which controls the Columbus research laboratory on the ISS. Cologne hosts both the European Astronaut Centre, which trains astronauts for their missions, and the headquarters of the German Aerospace Center, which developed the High Resolution Stereo Camera for the ESA's Mars Express mission, tasked with searching for traces of water and signs of life on the Red Planet. The country is a world leader in Earth observation and has produced two state-of-the-art radar satellites—TerraSAR-X and TanDEM-X—which provide highly accurate 3D images of the planet.

In 2021, the German military announced it was setting up a new space unit with the splendid name Weltraumkommando der Bundeswehr, which sadly translates only as Armed Forces Space Command. Its emphasis is on space as a defensive domain, focusing on space situational awareness and protecting German military and civilian satellites. When Defense Minister Annegret Kramp-Karrenbauer opened the base in Uedem, near the Dutch border, she noted how dependent Germany is on these satellites, "without which nothing works."

The UK, Europe's other major space power, remained a member of the ESA after Brexit, but with "third country" status. This ended involvement in Galileo and the ENGOS sat-nav system, and the UK was barred from Galileo's encrypted services used for defense and critical national infrastructure. In 2022, when the UK company Inmarsat began testing a possible replacement, both the ESA and the EU Agency for the Space Programme cooperated to ensure there is no interference between the services. The UK's divorce from the EU was acrimonious, but within the space community most people on both sides were saddened at the breakup and there remains goodwill to forge a new relationship.

The UK's story is very different from that of France. During the first few years of the Cold War it lacked the capacity to build its own satellites and rockets, and even now relies on the US for military satellite imagery. In return it hosts facilities for America's National Security Agency. However, in the 1960s, it did become only the third country, after the US and USSR, to have a secure military communication satellite system.

The breakup of the British Empire had left the UK with forces and intelligence centers all over the world that couldn't communicate with one another. Its satellite Skynet 1A was launched in 1969 from Cape Kennedy by an American rocket; others followed, and within a few years military communications in bases from London to Singapore were linked. Britain might have been retreating from empire, but it ensured that there was still concrete it could use around the globe.

The satellites were placed in geostationary orbit so that most of the UK's military and intelligence assets abroad were covered. There were gaps, but nowhere the British expected to have to fight. However, the enemy had the temerity to be in the "wrong" place—the South Atlantic. In 1982, Argentina invaded the Falkland Islands, which were outside Skynet's footprint. So when the Royal Navy task force arrived to persuade the Argentine Army to go away, secure communications between the military and the UK were . . . difficult. It helps to have friends. Special Air Service (SAS) teams made use of portable secure American Delta Force terminals they procured and the American Defense Satellite Communications System relayed messages to London. Without help from the US, the outcome of the Falklands War might have been different. It was enough of a scare for the UK to invest in the next generation of Skynet. The system went on to allow the British military to communicate independently during its operations in the Balkans, Iraq, and Afghanistan. Skynet 6A is currently being built by Airbus, scheduled to launch in 2025 and designed to withstand attacks from high-powered lasers.

Skynet 6A will come under the jurisdiction of UK Space Command, which was created in 2021 and is based at RAF High Wycombe in southern England. This new branch of the UK Armed Forces has been set up almost without anyone outside the military noticing. Nevertheless, it shows that the political and security elite have accepted that after decades of ignoring the space aspects of international relations and warfare (other than with Skynet), it now takes them seriously. The government has said it intends the UK to be a "meaningful actor" in the domain.

UK Space Command says its role is to "protect and defend UK and allied interests in space and control all the UK's defence space capabilities." However, it doesn't mention anything about going on the offensive, or about anti-satellite capabilities and plans. Its commander, Air Vice-Marshal Paul Godfrey, says: "Ultimately, we are doing space for defence. One of our goals is to protect and defend our assets in and through space . . . You wouldn't have an aircraft carrier floating around without any protection or understanding of what's going in that environment." They are also focusing on satellite communication for military operations; for example, those of the UK's special forces, such as the SAS and Special Boat Service (SBS), who could be helped by knowing when the opposition has surveillance capabilities above them, and when UK Space Command can see them and provide support. Some modern satellites are so good they can see through cloud cover and darkness, which previously provided strategic cover. "We can enhance their capabilities if our brothers and sisters in the other services know when they might be compromised," Godfrey explained. "Given how good satellites are these days, night and bad weather won't help them and so they might do things differently."

The UK may be one of only two serious military powers in Europe (alongside France), but in space terms it is far behind China, the US, Russia, Japan, France, the UAE, and others. Its Ministry of Defence is on a learning curve, and not all departments in it appear to fully grasp that twenty-first-century power politics and warfare are inextricably linked to space. A UK intelligence source specializing in space tech said: "In terms of tech we are cutting-edge and are concentrating on low Earth, that's where the investment is going, but overall, we are not up there with the big boys."

To help close the gap BAE Systems is working on building and launching a cluster of satellites into orbit that collect high-resolution visual images of Earth's surface, as well as radar and radio frequency information, even at night and in bad weather. Known as Azalea, the satellites' sensors can be reconfigured in space to suit the task they are given. Machine learning equipment on board the craft then analyzes the data, identifies activity of interest, and delivers it over secure channels to customers, most of whom are expected to be military. There's also a new spaceport in Cornwall, which, due to having the second-longest runway in the UK, will allow the very first plane-launched rockets on European soil. It's a serious step forward in the UK's capabilities.

The presence of the UK, France, Italy, and Germany inside the ESA makes it a powerful entity in space terms. It was the first of the space blocs. The second formal one to emerge was the Asia-Pacific Space Cooperation Organization (APSCO), which was formed in 2008 by China, Bangladesh, Iran, Mongolia, Pakistan, Peru, Thailand, and Turkey, with its headquarters in Beijing. It is modeled on the ESA and has a permanent council and secretariat. In a region plagued by earthquakes, and amid concerns about climate change, it makes sense to cooperate on developing satellites and sharing information. However, China calls the shots. The main aim of the organization appears to be to expand the footprint of China's BeiDou navigation system as part of the bid to overtake America's GPS as the dominant positioning tool.

The huge power of China is the focus of much space development and collaboration in the Indo-Pacific region. The divisions are shown by the existence of another grouping—the Asia-Pacific Regional Space Agency Forum. It was set up under Japanese leadership earlier than the Chinese-led APSCO but is a less formal institution. As it says, it is a "forum" not an "organization"—essentially a talking shop, but the talking is done mostly by countries less than friendly toward China, such as Japan and Vietnam.

Like the EU, Japan is a "civilian space power," but as tensions rise in East Asia it is finding it difficult to resist investing in military space equipment. However, its tentative steps toward this are closely linked with its continuing reliance on the US for military-grade satellite intelligence.

At the civilian level, Japan has an impressive space history, and an ambitious lunar program. It is one of only a handful of countries with its own launch ability. It sent its first satellite into space in 1970, and by 1990 had successfully orbited the Moon in an uncrewed craft. The state-run Japan Aerospace Exploration Agency has developed the Smart Lander for Investigating Moon, designed to allow landing craft to touch down within 328 feet of a proposed target area. As a signatory to the Artemis Accords, Japan will help with the Gateway Moon-orbiting station, and hopes to have a Japanese astronaut on the Moon via the 2028, 2029, or 2030 missions.

Japanese private enterprise is also getting involved. In December 2022, a SpaceX rocket carried a lunar lander known as M1 into space en route to the Moon. M1 was made by the small Tokyo-based company ispace, which hopes to win contracts from state agencies and commercial clients that want equipment delivered to the Moon or to map its surface looking for natural resources. M1's payload included the Rashid lunar rover from the UAE; a solid-state battery from Japan's NGK Spark Plug Company to test its resistance to cold; an AI flight computer from Canadian company Mission Control Space Services; and AI-enabled 360-degree cameras from another Canadian company, Canadensys Aerospace, which, among other tasks, were on board to film the UAE's rover.

The ispace company history makes for interesting reading. It began as one of the competitors trying to win the $20 million prize offered by Google in 2017 to the team behind the first private spacecraft to land on the Moon, travel 1,640 feet, and send back video. Known then as Hakuto, it focused on developing a rover but would have had to rely on a competing team from India to get it to the surface. The compromise was that both rovers would then have a 1,640-foot race. Admit it—you'd have watched that. Unfortunately, none of the teams were ready by the March 31, 2018, deadline and the prize went unclaimed.

Japan has been slowly but steadily rearming after decades as a de facto pacifist country. Its conventional forces are now equipped with offensive equipment, but when it comes to space Tokyo has retained a defensive posture. As a world leader in technology, and with a strong industrial base, its dependence on space-based communication systems means the economy is vulnerable if its satellites are disabled. Hence it has invested strongly in technology to track and dispose of space debris. Some of the tracking responsibility comes under the domain of the Space Operations Squadron (SOS), which works within the Air Self-Defense Force. It also monitors potentially hostile foreign satellites, but it's unlikely Tokyo will follow China, the US, and others in developing offensive space weapons.

The same is true of its neighbor, South Korea, but that country's technological brilliance means it too will grow as a space power. Seoul signaled its arrival as a "player" when it sent a lunar orbiter around the Moon in 2022 to study its chemical composition and magnetic fields. However, the country's current limitations were shown by the fact that the craft was lifted into space on top of a SpaceX rocket launched from Cape Canaveral.

Next door, North Korea has its own launch capability, usually from the Sohae Satellite Launching Station on the Yellow Sea. It has had limited success. Between 2012 and 2022, it tried to launch satellites five times but only succeeded twice, and it's uncertain if either satellite then functioned correctly. In late December 2022, Pyongyang said it had successfully put a satellite in space, and published images that included Seoul to back up its claim. This capability would mean North Korea does not need to rely fully on China for its surveillance intelligence. Even if its claim is correct, however, the extent of North Korea's satellite coverage remains limited. Pyongyang's neighbors, and the US, suspect that its satellite launches are more to do with testing its ability to fire possibly nuclear-tipped intercontinental ballistic missiles. North Korea's capacity to do so is unknown, but given what it has launched it seems quite possible it could hit another country's satellite with a direct-assault missile.

Another major player in the Indo-Pacific region, India cooperates closely with Japan and South Korea on civilian projects, but its space program is largely driven by the desire not to be left behind militarily by its great rival, China. India's main security concerns are centered on the Indian Ocean, where China now permanently stations warships, and the joint border with China in the Himalayas, which has seen armed clashes in recent years.

Indian space power has momentum but is growing too slowly for it to become a major military player this side of 2040. In 2019, New Delhi set up the Defence Space Agency but stopped short of establishing a full-blown space command, which is what the chiefs of staff had wanted. India has a military satellite system and a regional civilian satellite footprint but cannot match Beijing's spending power to develop a full-blown sovereign global navigation system.

Also in 2019, India successfully tested an anti-satellite weapon. China's 2007 test had shown New Delhi the direction of future space defense and how far behind it was. Successive governments had worked to strengthen global governance of outer space to prevent its militarization, but by 2019, India concluded it could not stand by as China and other powers advanced. It was a big decision. New Delhi had long criticized others for militarizing space. The test put it on the space military map. It has also been testing the waters in cooperating on space policy with its partners in the Quad, the Quadrilateral Security Dialogue (Japan, Australia, and the US)—quite a move from the once and possible future leader of the world's nonaligned movement. As so often, regional rivalries are a driver. New Delhi knows that China's expertise in space military activity will have a favorable knock-on effect for China's ally—and India's nemesis—Pakistan.

India is far more comfortable dealing with the civilian side of space. In 2008, its Chandrayaan-1 lunar orbiter discovered the probability of water on the Moon, including large water ice deposits at its lunar poles. This was among the factors that sparked the current global interest in building Moon bases. India cannot afford to build its own base or space station, but it has yet to sign up to the Artemis program. However, its commercial space industries are growing, and the Defence Space Agency has already provided satellite launch capability from its main launch site on the east coast near Chennai for dozens of other countries, including Indonesia, Malaysia, Turkey, Switzerland, Latvia, and Mexico.

Australia is a Quad partner of India, and much of its military thinking is also about China. However, unlike India, it has nothing with which to defend itself against a potential Chinese kinetic attack on any of the few satellites it possesses. It might be an enormous country, but it is currently small in terms of space capabilities. That is now set to change—Australia intends to become a middle-ranking space power by 2030, as befits its status as a current middle-ranking land and sea power.

Australia's position in the Southern Hemisphere attracted a friend looking for a safe spot to establish intelligence-gathering and space-tracking stations: the United States. Bases in Australia can be situated in remote locations, which helps keep them secure and means there is almost no radio frequency interference. They can look at parts of space that cannot be seen in the Northern Hemisphere and are well placed to monitor China's space-launch trajectories and geosynchronous orbits. In 1961, Australia signed an agreement with the US that established several such bases across the country. Some were used to track the rockets of American space missions, including the Moon landing of 1969. The best known is the Pine Gap facility, which, if it wasn't relatively near Alice Springs in the Northern Territory, would be described as being in the middle of nowhere.

Pine Gap is, arguably, the most significant American intelligence-gathering facility outside the US and one of the strongest bonds tying the two countries together in a relationship of mutual trust. Australia is under the US's extended nuclear umbrella and that requires a contribution to its effectiveness. The base was opened in 1970, but only named Joint Defence Facility Pine Gap in 1988. The word "Joint" reflected a change in how it operated. Australian defense officials were given senior management positions, including deputy chief of facility, and it became a mantra that all activities at Pine Gap "take place with the full knowledge and concurrence of the Australian government."

In 2013, the then Australian defense minister Stephen Smith

repeated this in an address to Parliament, while also stating that the alliance was now expanding "cooperation in the modern areas of cyber, satellite communications and space." Among the facilities at Pine Gap is the Relay Ground Station for America's Space-based Infrared System (SBIS), which gives early warning of ballistic missile launches. There are more nuclear-armed countries in the Indo-Pacific than any other region—China, North Korea, Pakistan, India, and the US—and so Australia's access to SBIS is a vital defense asset.

In 2022, Canberra set up its Defence Space Command within the Royal Australian Air Force. It was a sign that the government recognizes this new domain of geopolitics and warfare and that it requires a level of sovereign autonomy there. This was reflected in a document released the same year, which talked about developing capabilities that "can be reconstituted if compromised and defended if under attack." This was a reference to building large numbers of small satellites that can be quickly replaced if destroyed in orbit. It's not specified how many would be military satellites, but it's probable some would at least be "dual use." The head of Defence Space Command, Air Vice-Marshal Catherine Roberts, accepts that Australia is "far behind" and needs to "accelerate the capability so that we can deal with the threats."

The need to create a space command was increased when Australia signed the AUKUS (Aus/UK/US) defense alliance in 2021. AUKUS is centered on a deal to supply nuclear-powered submarines to Australia, but within it is the understanding that the three need to cooperate in space. The Americans had Space Force, the British had Space Command, and so Australia's Defence Space Command was set up within months of the pact being signed.

Commercially, Australia is late to the game—its civilian Space

Agency was only established in 2018. It is small but focused and has the ambition to grow the domestic commercial industry from 10,000 jobs and a value of A$3.9 billion to 30,000 jobs and a value of A$12 billion by 2030. That is a bold target, but at least they have launched. Australia's lack of its own significant satellite system means it is currently dependent on other countries for weather forecasting, and for monitoring natural disasters from volcanoes to bush fires. It is reliant on data from Japan, China, the ESA, and the US. A ten-year plan to rectify this should see an Australian constellation of satellites dedicated to weather, communication, and the military.

When it comes to space, the relationships among the Indo-Pacific countries reflect the politics and economics of the region. China seeks to dominate, and it set up the Asia-Pacific Space Cooperation Organization to undermine Japanese influence in the sector. It has had some success with this policy by pulling in developing countries and covering part of the costs of involvement. Japan and India have responded by increasing their military capabilities in space and stepping up cooperation with each other and Australia. China is the biggest player but has few friends, even among the APSCO members, whereas almost all other countries in the area have one thing in common—anxiety about being overwhelmed by China's weight.

This split means there is little chance of the entire region uniting in a single organization. Happily, it still leaves room for a lot of collaboration on scientific and commercial projects, but militarily the future looks bloc-based.

In the Middle East there are several rising space powers, and future alliances are yet to be decided.

Israel, one of the smallest countries on Earth, established a space agency as early as 1983 under the control of the Ministry of Science and Technology, and within six years had launched its first satellite. In the previous decade, the country had been shocked when its military warning system had failed to spot the surprise invasion by Egyptian and Syrian forces in the Yom Kippur War. The government concluded it needed its own satellite reconnaissance capability.

It was starting from scratch on the satellite technology, but it benefited from its knowledge of rocketry, having developed ballistic missiles in the 1960s in cooperation with France. In the 1980s, its Jericho 2 rockets, designed to carry nuclear weapons, were adapted to become the Shavit space-launch vehicle. Israel now has a constellation of reconnaissance and communications satellites.

Most countries launch space rockets eastward, which as we've seen gives them a boost from Earth's rotational speed, but the Shavit launches due west—against the planet's spin. This "retrograde" launch is to ensure the rockets fly over the Mediterranean Sea and not over Israel and then neighboring Arab countries, some of which remain hostile to it. This is done for the protection of populations, and because Israel does not want its Arab neighbors mistaking a space launch for a missile attack.

The Shavit route takes the rocket up and straight over the Mediterranean, after which it tries to thread the needle of the Gibraltar Strait and continue upward over the Atlantic. The westward direction requires more fuel to get out of the atmosphere, which reduces the amount of weight the rocket can carry by 30 percent. This is a disadvantage, but Israel has partially turned it into a positive. Just as the country's security challenges led to the development of its space capabilities, so retrograde launches have spurred breakthroughs in miniaturizing technology and developing lighter satellites that still deliver high-resolution images and secure communications. The smaller the satellite, the more you can launch on one rocket, thus making the enterprise more cost-effective.

Israel has developed formation-flying nanosatellites and is working on the ULTRASAT (Ultraviolet Transient Astronomy Satellite) space telescope, to be launched into orbit in 2026. Its National Knowledge Center on Near Earth Objects aims to map objects that could threaten Earth and find a solution to deal with them, and the Israel Cosmic Ray Center on Mount Hermon monitors potentially dangerous space phenomena such as solar storms.

There's also the ambition for Israel to return to the Moon. Yes, return. The privately funded company SpaceIL sent the Beresheet spacecraft to the Moon in 2019. As it slowed down above the Sea of Serenity there was a hardware malfunction, causing it to crash-land. The craft is still there, and still inside it is a miniature copy of the Hebrew Bible, the first word of which is *Beresheet*, meaning "Genesis" or "in the beginning."

It was just the start. A year after the crash Israel and the UAE signed the Abraham Accords, normalizing their relations. Both are world leaders in space technology, and so there was a logic to the 2022 announcement that the Beresheet 2 mission would be a joint project between them, although SpaceIL will take the lead.

Scheduled to launch in 2025, the plan is for a mothership to orbit the Moon and then release two lunar landers, one on the surface facing Earth, the other on the far side—a region to which only China has so far ventured. If successful, it will be the first double landing, and the two landers would be the smallest craft to get there, each weighing just 265 pounds, including fuel. The mothership would then orbit for five years, sending data home, including information about climate change, desertification, and water supplies—subjects of interest to both countries.

In 2019, the plaque on the side of Beresheet 1 read: "Small country, big dreams." The phrase could also apply to the UAE, which has the most ambitious space program in the Middle East.

The tiny, energy-rich Arab country only launched its first observation satellite in 2009 (from Kazakhstan) and didn't have a space agency until 2014. But on February 9, 2021, its Hope spacecraft arrived in orbit around Mars to study its atmosphere, making the UAE only the fifth entity in history to reach the Red Planet after the US, the Soviet Union, the ESA, and India. China became the sixth when its Tianwen-1 craft showed up just twenty-four hours later.

The chair of the UAE Space Agency, Sarah Al Amiri, is not content with this stunning achievement. Her team is now working on a mission to fly a craft 2.2 billion miles, taking in a flyby of Venus, followed by landing on an asteroid. The launch target is 2028, and the landing is scheduled for 2033. The agency was established as part of the UAE's broad economic diversification away from reliance on oil and gas revenues. This fits with the ambition of turning the Emirates into a hub for advanced technology. It can already build its own satellites and is developing a small satellite constellation called Sirb—Arabic for a flock of birds.

The UAE, like Israel, is a signatory to the Artemis Accords. This does not prohibit cooperation with other parties, but eyebrows have been raised by how far the Emirates is allowing China into its space industry. The phone company Huawei is deep inside the country, having built its 5G network. Assurances that it has not included any back doors through which it might tap into security information have not allayed the anxieties of the country's Western friends. In 2021, the US suspended a deal to sell the UAE fifty F-35 jets, citing security concerns. The fighter jets wouldn't have been using the Huawei 5G technology, but the system's ground stations and communications towers could easily gain insights into how America's latest-generation fighter jet operates. In early 2023, the two sides were still discussing how to resolve the issue.

The UAE intended to fly its Rashid 2 rover on the Chinese mission to the Moon in 2026, and possibly operate near one of the proposed landing sites for the Artemis 3 crewed mission. However, in March 2023, US technology restrictions preventing collaboration with China grounded the plan as Rashid 2 has American-made components.

The one other country in the wider Middle East with its own launch capability is Iran. In 1999, Tehran announced the ambition to build satellites, along with the rockets required to place them in orbit. However, the rocket side of the equation was mostly used as cover to develop long-range missiles.

The Iranian Space Agency comes under the jurisdiction of the Ministry of Communications and Information Technology, but the companies making space rockets also build missiles and are subsidiaries of the Ministry of Defense. The country's strongest military force, the Islamic Revolutionary Guard Corps (IRGC), operates its own space program and reports directly to the Supreme Leader and not the president. In 2020, it launched its own explicitly military satellite, adding to the handful of domestically built satellites Iran operates. A second reconnaissance satellite followed in 2022.

So Iran can build, launch, and operate, but it can't yet do those things very well. The rockets frequently fail, and the satellites are usually of low quality, have a short life span, and are confined to low Earth orbit. Nevertheless, Iranian scientists are learning and improving, and are ambitious, although a pledge to put a man into space by 2025 seems a bit of a stretch. In 2013, Iran had set a deadline of 2018. The then president, Mahmoud Ahmadinejad, heroically volunteered to be the first Iranian astronaut, and said he was willing to sacrifice his life for Iran's ambitious space program. Luckily for him, the project never really took off. Now, with the 2025 deadline looming, 2032 has been deemed a more realistic date with destiny.

When President Ebrahim Raisi came to power in 2021, his administration lamented the "sorrowful" state of the space program and vowed to invigorate it. The head of the space agency was fired, and a commitment was made to getting a satellite high up into geostationary orbit within five years. Iran, it was declared, must be the preeminent space power in the Middle East. The number of satellite launches for low Earth orbit has increased and there are plans for a new launch center to be built in the port city of Chabahar. Located in the southeast, it is close to the equator and so rockets can travel east over the Indian Ocean after liftoff.

Iran tries to limit the space advantages developed by potential adversaries such as Israel and the United States. Theoretically it could try to adapt one of its medium-range ballistic missiles into a satellite killer, but the precision required to hit something 186 miles above Earth and traveling at 4.8 miles a second is currently way beyond its capabilities. Jamming and spoofing satellites is cheaper and easier, and Tehran has experience doing it. For years it has been jamming dozens of Persian-language broadcasts beamed into Iran. Cable-based internet services are severely censored, and so millions of ordinary Iranians turn to satellites for information, meaning the government is in a constant battle to find and block the signals coming from outside the country.

Iran uses space for the same civilian and military purposes as many other countries and, like most, has disguised the military aspects. Because of Iran's efforts to obtain nuclear weapons (which it denies), many advanced powers are nervous about its space program and seek to disrupt it. However, much of the less technologically capable world agrees with Tehran's view that space cannot belong to the exclusive club of those who got there first and must be open to all for scientific, economic, and—yes—even military advancement.

The African countries certainly agree. Many of them have their own national space agencies, among them South Africa, Nigeria, Kenya, Botswana, and Rwanda. Few have serious short-term ambitions to participate in space exploration, but they argue that any legal framework around activities in space must be a global effort. Most of them haven't taken a firm side in the US-China race for space and will work with whichever nation offers the best terms for accelerating their own space industries. For example, Nigeria's first two communication satellites were launched by China, but in 2022, it signed the US-led Artemis Accords along with Rwanda. Russia has launched the most satellites for African countries, followed by France, the US, China, India, and Japan.

The African Union (AU) lists developing a continent-wide space strategy as one of the key programs in its Agenda 2063, a long-term framework for raising living standards for the 1.2 billion (and rapidly rising) population. It recognizes that Africa cannot afford to remain a net importer of space technology and supports the fast-growing industry of space-focused start-ups. However, despite passing a resolution to establish an African Space Agency in 2017, and choosing Egypt to host its headquarters, it has made little progress. Instead, individual countries are forging ahead.

Although many countries have their own national space agencies, there are no launch facilities on the continent. During the apartheid years, South Africa was a nuclear power and had the capacity to send rockets into space from the Denel Overberg Test Range on the coast east of Cape Town. It launched test flights for Israel's Jericho 2 missiles, as well as three South African rockets that reached suborbital trajectories in the late 1980s. But that changed in 1989 when F. W. de Klerk came to power seeking to end the apartheid era and ordered the nuclear program to be terminated. In 1991, South Africa signed the Nuclear Non-Proliferation Treaty and as part of the process the long-range-missile launch pad at Overberg was taken apart. Since then, no African country has had its own launch capability.

That may change. In early 2023, Djibouti signed a memorandum of understanding with China's Hong Kong Aerospace Technology Group (HKATG) to build a spaceport there. The plan is for the tiny Horn of Africa country to provide a minimum of 3.8 square miles of land and a thirty-five-year lease, at the end of which the infrastructure would be handed over to the Djibouti government. The $1 billion project would see the building of port facilities and roads to help transport Chinese aerospace equipment to the site where seven satellite launchpads and three rocket-testing pads are envisaged.

If the project goes ahead, China gets a spaceport in a key location in Africa—Djibouti is close to the equator, which lowers launch costs. It is also home to a Chinese naval base, giving China access to the Red Sea, right at the choke point where the sea narrows before opening into the Gulf of Aden. Adding a space facility to its portfolio in such a strategically important place gives Beijing serious leverage in the region. Djibouti gets prestige, inward investment into a high-tech industry, and, a few decades down the line, a spaceport.

In the immediate future, satellites are going to be a major area of growth for many African countries. Most African economies are heavily dependent on agriculture and as such vulnerable to the effects of climate change. Since the continent's first satellite went into orbit in 1998, more than forty have been sent up and the launch rate is increasing. The first was Egypt's Nilesat 101, whose role was to deliver multimedia services to five million homes, but now most satellites are designed to monitor the environment. The data can be used to map changes in the size of forests and lakes and can act as an early-warning system of impending problems. It can also boost food production. The University of Ghana has partnered with the Rainforest Alliance and other groups on the SAT4Farming project, which helps tens of thousands of the country's cocoa farmers to increase crops and income by providing information on individual plots of land.

South Africa builds its own satellites and in 2022 used SpaceX to take three nanosatellites, designed and built in Cape Town, into orbit. The three machines, each measuring just 8 inches × 4 inches × 4 inches, are part of its Maritime Domain Awareness Satellite constellation, detecting and identifying ships off the country's coast. South Africa's Exclusive Economic Zone (EEZ) extends 200 nautical miles from its shore, and as it has such a long coastline the EEZ is larger than its land area. Its nanosatellites now allow it to control its territory with far more accuracy than in previous decades.

Nigeria also has its own satellites, which have helped the government to monitor the Boko Haram insurgency in the north. However, the limitations in coverage were exposed in 2021 after another mass kidnapping of schoolgirls took place. The head of the National

Space Research and Development Agency admitted that the agency couldn't track the movement of the group that had taken the girls because its high-resolution imaging satellite was "not static [above] where the insurgency is taking place." More satellites will give greater coverage and will assist in Nigeria's long tradition of sending peace-keeping forces to the continent's war zones.

The other major focus in Africa is astronomy. The relatively clear night skies there have attracted serious investment from outside companies and interest from academics. South Africa, Ethiopia, Egypt, Nigeria, Namibia, Mauritius, and Ghana all host major astronomical observatories, and there is a growing industry in astrotourism for keen amateur astronomers.

Southern Africa is particularly well positioned for both visual and radio astronomy. There are vast areas of mostly uninhabited land with "quiet" radio zones, clear skies, and direct sight of the Milky Way. This view is why one of the world's largest radio telescopes is situated in the Northern Cape in South Africa. The MeerKAT telescope was funded by the South African government and built at a cost of $330 million over ten years. It consists of sixty-four satellite dishes, each sixty-five feet tall.

Since its launch in 2018, MeerKAT has had a string of successes, including finding giant galaxies previously hidden despite being twenty-two times the size of the Milky Way. Over the next few years it will be incorporated into the Square Kilometre Array (SKA), an international project of almost 200 linked satellite dishes and 131,000 antennae in South Africa and Australia financed by more than a dozen countries including India, China, Italy, and Portugal. When finished, circa 2030, it will be the largest scientific structure in the world—although it is spread over more than 93 miles, if you put the dishes and antennae all together they would cover about 1 square kilometer (.39 miles), hence the name.

SKA will be able to see through the clouds of cosmic dust that obscure optical space telescopes and is expected to revolutionize our knowledge. It's claimed to be so sensitive it could receive radar signals from an airport on a planet trillions of miles away—should such a thing exist. It is an example of how cooperation between countries and companies benefits everyone. We've seen numerous others over the decades of the Space Age, and despite the polarization of the world there are still many scientific and commercial endeavors under way. However, when it comes to hard politics, we're back to hard realities.

None of the countries mentioned above, nor other potential players such as Brazil, Turkey, and Indonesia, look anywhere near ready to challenge the status of the Big Three space powers. Elsewhere we find the seven-member Latin American and Caribbean Space Agency (LACA), which was created in 2020 and, as with the African Space Agency, is focused on development. There's also the Arab Space Cooperation Group, formed in 2019, but there's been little communication. The eleven member countries do meet annually; however, to date most activity happens at the state level. Collaboration in these space blocs, while simultaneously building sovereign satellite capacity, makes sense for most countries, although the progress of the African Space Agency shows the pitfalls if the bloc makes slow progress.

Apart from the ESA, the two blocs that really count in geo- and astropolitical terms are the US-led Artemis Accords and the Sino-

Russian lunar agreement. All three are attempting to shape norms of behavior in space and international law. In broad-brush terms, the ESA is closer to the American view than the Sino-Russian. Other countries must weigh not just how they feel about a particular space issue, but how siding with one or the other bloc will affect their relationship with it. As the sector grows in economic and military importance, so will the pressure to take a view. As on Earth, so in space.

PART III

Future Past

Illustration showing a military satellite shooting a laser in space.

9

SPACE WARS

Two things are infinite: the universe and human stupidity;
and I'm not sure about the universe.
—Albert Einstein

Each time humanity has ventured into a new domain it has brought war with it. Shipbuilding resulted in warships. Airplanes brought fighter jets and bombers. Space is no different and the potential battlefield is beginning to take shape.

We know there's a lack of meaningful frameworks to guide peaceful operations in space; that more and more countries are getting involved; and that tensions are already surfacing around hot spots, from Lagrange points to Moon bases. If we are heading for space-based conflict, what might that look like?

Conceptually, some astropolitical writers argue that space warfare should be framed as a fight for celestial lines of communication—just as countries have contested Earth's sea lanes, and the communication and trade that go with them, so they will contest orbital lanes. Others, such as space warfare expert Dr. Bleddyn Bowen, refer to the orbital lanes as a "cosmic coastline," given that land powers can project power into space, dominating the area above them as they would the seas around their coasts. Thinking of space as "high ground"—as the place that needs to be occupied in order to command the land/battleground below—can also be useful to the layperson. Not everyone agrees with the term. Dr. Bowen feels it is misused, that it implies that space assets must be defended, and prefers to refer to space simply as "a place where there is some advantage to be gained."

But no matter how the issue is framed, most analysts agree that in the near term no one power can dominate space, and that—for now—even being the most powerful spacefaring nation will not guarantee command of Earth. However, they also generally agree that as the importance of space grows both militarily and economically, so will the degree of competition. And because theoretically one power could eventually gain control, all the major countries are investing in the sector, ensuring they will not be squeezed out, while the second-tier powers are trying to reduce their reliance on, or dominance by, the Big Three.

The pieces needed for what might become our first "space wars" are already in place.

For this decade at least, a war in space would primarily be about war on Earth. Given that technologically advanced powers now rely so much on space, the domain is central to modern military thinking. Without satellites, the commanders don't know where to position their aircraft carriers, long-range missiles, and troops. Nor do they know precisely where the enemy is.

Professor Everett Dolman suggests that, in the near future, any space conflict is most likely to stem from the tensions in the Asia-Pacific region involving China, Taiwan, India, Japan, and the United

States: "The ability of the US to project military power today is based almost entirely on space support. This includes precision guidance, intelligence and surveillance, and the political will to act that comes from the illusion of perfect knowledge of enemy deployments and intentions. It is therefore a tremendous advantage for China to take out US space support prior to initiating a terrestrial military action that would be opposed by the US."

It's not inevitable, and there are many restraining influences, but this has also been true in the past when, via miscalculations and misunderstandings, countries have blundered into wars. Countries have also embarked on wars of choice. The below scenario, based on a war of choice, is just one possibility for how we might see space playing a bigger role in our conflicts on Earth.

———

May 2, 2030, 3:09 a.m. Cheyenne Mountain Air Force Station, Colorado.

A Specialist Four (Spc4/E-4) Space Systems Operator on the night shift notices two Chinese satellites have moved closer to an American satellite that has oversight of the Taiwan Strait. She's a relatively junior "Guardian," as Space Force members are called, but given the Chinese military buildup on its coast, knows that this needs to go up the ranks quickly.

The PLA has spent the previous three months moving ships, troops, and landing craft into place on the coast, signaling a possible attack on Taiwan. The Americans are puzzled. The positioning of the buildup points to an invasion across the strait, but the number of landing craft is nowhere near what is required for an amphibious assault.

May 2, 7:24 a.m. The Chinese satellites have edged even closer, and the Spc4's initial report is in the White House. A brisk diplomatic message is sent to Beijing: "You're tailgating. Back off." The response comes back the same day. China insists its satellites have no ill intent and quotes from the 2002 UN Treaties and Principles on Outer Space: "there shall be free access to all areas of celestial bodies." For good measure, the People's Republic reminds Washington of the crisis of 2028 when the US "inspected" a Chinese satellite at close range.

Tensions remain high through most of May, especially after the US launches two small "bodyguard" satellites that maneuver into position between the Chinese and American machines. A week later, in a show of solidarity, the UK does the same.

June 1. The story has fallen out of the headlines—nothing has happened, and anyway, it's monsoon season in the strait, hardly invasion weather.

September 4. The waters are calm, but diplomatic tensions are about to head into rough seas.

September 12, 9:20 a.m. An Australian satellite, linked to the Five Eyes intelligence-gathering network, mysteriously falls out of orbit and descends into the atmosphere before burning up. Another Chinese satellite slowly maneuvers closer to a US satellite, this one part of the command-and-control system for the US's nuclear deterrent. Washington ups its alert level. Potentially endangering its ability to monitor the strait is one thing, having the capacity to mess with its nuclear deterrent is another. If part of the early-warning system goes dark, the US will be susceptible to a surprise nuclear attack.

The Americans demand an emergency UN Security Council meeting at which they propose "zones of clearance" for satellites by which other countries cannot venture within a certain range. Nothing comes from the meeting or the demand. Beijing repeats that it adheres to the rule of law, this time pointing to the Outer Space Treaty, which states that the domain "is not subject to national appropriation by claim of sovereignty."

September 19, 7:41 p.m. As Chinese ships begin rehearsing loading and unloading troops, Washington moves an aircraft carrier fleet from Tokyo Bay with orders to rendezvous with a Japanese carrier off Okinawa, an hour's flying time from Taiwan. The British dispatch the *Queen Elizabeth* carrier fleet from Portsmouth, and Australia's new nuclear-powered submarines head toward the Philippine Sea. India and South Korea call for calm.

October 3, 4:00 a.m. (Pacific Time). It happens. But not quite what the Americans had feared.

The Chinese fleet leaves from ports along the coast with air cover. Twenty minutes later the two satellites "tailgating" the American satellite responsible for the strait dazzle its cameras, rendering it blind. At the same time, American, Japanese, and Australian satellites across the region are either jammed or spoofed with confusing signals sent by the Chinese. The moment this happens, the "invasion fleet" begins to return to port, but the Chinese jets covering them head straight down along the coast to the Taiwanese-controlled Kinmen Islands, just 1.9 miles from the Chinese mainland.

The PLA had been beaten back from there in 1949 and failed to overcome the islands again in 1958, but this time the battle is over almost as soon as it begins. Taiwan has let the garrison run down from 50,000 troops in 2000 to just 3,000 during the 2020s. It is relying on the latest version of the unmanned self-firing short-range weapons system on the Wuchiu islands, first deployed in 2022 to deter a third attempt. But Chinese electronic warfare operators based near Mumian, on Hainan Island, are inside the system. As the small number of special forces storm ashore after crossing the short distance in small, fast boats, the guns are mostly silent. Besides, the few still operating are pointing the wrong way. The special forces aren't the biggest problem. With the Taiwanese Air Force patrolling the main island 116 miles away, unaware of the real target, 20,000 Chinese paratroopers drop in on Kinmen as their air force gives them complete cover. Thirty percent of the depleted garrison's defenses are destroyed in just the first wave of attacks. It's a fait accompli—the surrender comes at 9:50 and 160,000 of the island's citizens are now under the control of the People's Republic of China.

Taiwan appeals to the Americans to join them in a counterattack. Washington declines and Taiwan knows it cannot go it alone. But the Americans are well aware there has to be some form of response.

October 4, 10:10 a.m. (PT). It takes a day, but two "bodyguard" satellites, equipped with small booster rockets, maneuver themselves above the Chinese satellites and use their robotic arms to push them down into the atmosphere, where they are destroyed. The Chinese are outraged, but what follows is even more dangerous.

12:55 p.m. (PT). The Americans target the Chinese machine closest to one of their nuclear command-and-control satellites, blowing it into several thousand pieces with a laser. They use the X-40A uncrewed spaceplane, an updated version of the reusable X-37B that had developed laser capacity in the early 2020s for peaceful purposes. The attack is at an angle, which means that most of the 4,000 bits of resulting debris go into deep space, but hundreds of small pieces remain in orbit, adding to the dangers already faced by astronauts from several countries, including China. To add insult to injury, another US satellite has been tracking a Chinese model used for naval communications. Over a period of twenty-four hours it moves in, grabs the satellite's antenna, and bends it 180 degrees. A cosmic fender bender.

Threats by Beijing to retaliate in kind come to nothing. Eventually the crisis subsides, but the fallout will last for years. The US, Japan, Australia, Indonesia, and the UK sign a defense pact with Taipei saying they will come to its assistance in the event of "an attack on the mainland." Missing is a guarantee to defend the other islands between Kinmen and the mainland. Also missing, despite the first acknowledged military action in space, is a treaty on Space Situational Awareness, which advocates nickname GOOMO: "Get Out of My Orbit."

———

And . . . back down to Earth. Any future scenario is purely theoretical, and the above is probably flawed, but most of the technology in it already exists. The U.S. Space Force has Space Systems Operators; France has developed bodyguard satellites that could carry weapons for "active defense" purposes; dazzling and spoofing equipment is already available; the Wuchiu self-firing artillery has been deployed by Taiwan; and there is an X-37 spaceplane.

While space can already be used to conduct war on Earth, for the foreseeable future warfare in space itself will be very slow. Satellites can already attack each other, but maneuvering any spacecraft requires unhurried, deliberate, and precise moves. Operators must calculate the intersection of different orbits to get a machine into a position where it can grab, ram, or even shoot at another, so changing the orbit of a satellite takes a lot of effort. And while they may be moving extremely quickly, faster than a speeding bullet, space is very,

very big. Take the area between low Earth orbit (starting at about 100 miles above us) and geostationary orbit (22,236 miles up). The volume between the two orbits is 190 times larger than the volume of Earth. That's a lot of space to cover.

If anyone makes a current real-time war film in space with one satellite chasing another, you're going to need a day off work and a lot of popcorn. And coffee. On the plus side, the chances of missing some action during a bathroom break are slim.

There are advantages and disadvantages to this slowness of movement. It gives potential adversaries time to contact each other and attempt to resolve what might be an approaching crisis. However, it also increases the risk of preemptive attacks. If a state sees a rival moving several satellites into what could become unbearable threatening positions, it may be tempted to attack what is sometimes called the kill chain—the infrastructure on Earth that supports the potential enemy's satellites. This might be done via cyberwarfare. Even if it were diplomatically handled—for example, by signaling that this was a proportional response to threatening behavior, and that no more attacks would be undertaken—the action could easily spark retaliation. The country that suffered the first strike could hit back by firing a direct-ascent anti-satellite weapon (ASAT) at one of the other side's satellites, then also claim proportionality. At which point anything could happen, from no further action to nuclear war.

ASATs are a constant threat for all satellites, but a major issue is the security of satellites that are crucial to the nuclear-armed countries' early-warning systems. Some of them warn about the launch of what might be a nuclear missile, while others (e.g., within the US's Advanced Extremely High Frequency network) are for use in communications in the aftermath of a nuclear strike. Each costs north of a billion dollars, is the size of a small house, and any sign of them being threatened makes those who own them very, very twitchy.

Future models will be more sophisticated and more expensive. The US is building its Next-Generation Overhead Persistent Infrared warning system, to be in service by 2030 and costing billions. These will be the size of houses, very tempting targets, and another example of the need for treaties, especially given the Space Situational Awareness (SSA) problem outlined in our war scenario above.

In the absence of agreement on such matters, and as competition heats up, the likelihood of conflict grows. We might not be there yet, but consider the following scenario, which might not be far off.

⸻

April 4, 2038, 5:10 (Moon Time). Artemis Integrated Moon Structure (AIMS).

The Japanese Lunar Watch shift has been tracking a Russian spacecraft since the previous day when it launched from the Plesetsk Cosmodrome, north of Moscow. Within a few minutes it had been clear that it was on course for the Russian Moon station 310 miles from the multinational AIMS base, but in the last hour the night shift operators notice that its trajectory has changed. Now it seems to be heading for a spot between the two. Then it changes course again. The operators do a few quick calculations and hit the alarm button.

In a clear violation of the Artemis Accords, the Russian landing craft is heading straight for the British base in the Peaks of Eternal Light region at the lunar south pole. But Russia is not an Artemis signatory, and Moscow has long argued that it is not bound by any of its articles, not least the self-styled "safety zones," such as the one the British declared near the Shackleton crater. This is prime lunar territory due to its huge resources of frozen water and methane inside the crater, which, unlike its peaks, is in perpetual shadow.

The alarm sounds at all four Artemis nation-state bases, British, American, Japanese, and UAE, but it's the Brits who have to move quickly. A robotic rover is driven out to block the landing track, and the station's air locks are double-secured. At 5:55 the Russian craft begins to glide in along the right-hand side of the landing track to avoid the rover and take advantage of the relatively flat ground. At 6:09 disaster strikes. A small, smooth boulder catches the craft's underside, causing a wing to tip and hit the surface. The force of the impact swings the body round 360 degrees and slightly to the left, and just 164 feet on it smashes into the rover and breaks in two.

When the British medical crew reach the front section, they find six dead cosmonauts. In the back the salvage team discovers two vehicles: one, a basic robotic building machine, the other a rover with drilling capacity. It looks as if Moscow intended to put "facts on the surface" and underline its opposition to "safety zones," a term Russia says is a smokescreen for lunar spheres of influence.

April 6, 8:38 (MT). At the emergency meeting of the UN Space Council, the British offer their "sincere condolences" for the tragic loss of life but say it was unfortunate that Russia had ignored the "safety zone." The Russians blame the UK for blocking the landing site and remind everyone that the 1979 Moon Agreement describes the Moon, and its resources, as the "common heritage of mankind." The Americans point out that the treaty has never been ratified. The Chinese keep quiet. Crisis over?

April 13, 5:12 (MT). A second attempt. This time the Russians announce they are on their way. Moscow tells the American North Link company it intends to land at its Bore base at the lunar north pole and begin drilling for rare-earth materials. North Link replies that this is in violation of its commercial rights, especially as it has spent a fortune proving exactly where the materials are. Washington warns Moscow that its duty to protect American citizens extends beyond Earth's boundaries and places Space Force on heightened alert.

As the Russian craft begins its descent, the landing track is blocked with three rovers and a looped warning is broadcast on the Russians' frequency. A minute out, the Americans in the forward operating bases (FOB) to the front, left, and right of the track begin to use a laser to dazzle the incoming vehicle from each direction, assuming the Russians will pull up and head back out. What happens next is unexpected. The Russian craft fires a directed-energy beam at the front FOB. It hits the machine firing the laser with such force that it explodes. The shrapnel punches a four-inch-wide hole in the side of the FOB, and small tears in the pressurized suit of one of the two laser operators. She dies well before the medics reach her after a desperate rescue mission.

The Russian craft has indeed pulled up and heads back to rendezvous with the Russian space station, but there's no time for an emergency UN meeting or statements of outrage—the Americans simply open fire. They hit an electro-optical sensor at Russia's Zelenchukskaya base in the North Caucasus using a cruise missile. Simultaneously they blow three Russian spy satellites out of low Earth orbit using ground-based direct-ascent missiles. Four commercial satellites are disabled with a cyberattack, causing most of the Russian cell phone system to go down along with the Moscow stock exchange. The cost to the Russian economy over the next eighteen hours is conservatively estimated at $760 million.

The attacks are calibrated. The targets are not directly linked to Russia's nuclear early-warning systems, and the Zelenchukskaya missile only killed three soldiers from the Third Army's space surveillance division. Russia's next move baffles analysts. Surely they got the message—Washington reacted with what it clearly thought was an appropriate response. Moscow could now either leave it at that and let the diplomatic channels take over or reply in a similar manner, meaning containable action. Instead, over a period of forty-eight hours, it maneuvers six killer satellites into position behind American craft linked to Washington's nuclear missile early-warning system and begins attacking them. Four are hit before Space Force takes out all six Russian craft using direct-ascent missiles. Part of the US early-warning system has gone dark, leading Washington to go to alert level DEFCON 2, reputedly for the first time since the Cuban Missile Crisis in 1962. Moscow follows suit, declaring "Elevated Degree of Combat Readiness"—one down from imminent nuclear action.

The Americans quickly replace their satellites with an emergency stock held next to their military space station, meaning both sides can now see the other readying their nuclear arsenals and moving troops and ships. The world holds its breath. And then, with the White House and the Kremlin each conducting meetings to consider a first strike, the Chinese pick up the phone.

And that's how nuclear war is avoided in 2038. Beijing hosts a trilateral summit and the Big Three agree to introduce several "confidence-building measures," including an agreement that all mining lasers deployed to the Moon can only point downward. On the surface, tensions die down. Everyone realizes that Mutual Assured Destruction (MAD) has been tested almost to a breaking point for the second time in less than a century. Just as the Cuban Missile Crisis did in 1962, this incident has concentrated minds, averting disaster. Third time around we might not be so lucky.

———

The most dangerous aspect in this scenario is the prospect of a nuclear-armed state's early-warning system going dark. The probability of a preemptive strike would rapidly escalate if a country could not find an explanation for why a potential adversary had blinded it.

There are several other dangers that are clear and present now or lurking very soon in the future.

For example, an ASAT exchange between India and Pakistan could drag in their allies—or worse, the two nuclear-armed countries could escalate their actions.

A rogue state could develop a killer satellite fleet in secret, launch it, and hold a country, or indeed the world, for ransom.

Another rogue state, bitter at being frozen out of an agreement to benefit from space exploration, could explode several huge nuclear bombs in low Earth orbit, frying most of the satellites and plunging the world into chaos.

Sci-fi? In 1962, the US launched a military project code-named Starfish Prime. They detonated a thermonuclear warhead 250 miles above the Pacific Ocean—just to see what would happen. The device was a hundred times more powerful than the one dropped on Hiroshima. Within seconds an electromagnetic pulse knocked out electricity in Hawaii, and from Hawaii to New Zealand the night sky lit up in a carnival of color, a man-made aurora. An artificial radiation belt formed around Earth, lasting a decade before dissipating. At least seven satellites were damaged or destroyed, including the Telstar communications model. "Oops," said the Americans. Or rather, as one scientist put it later, "To our great surprise and dismay, it developed that Starfish added significantly to the electrons in the Van Allen belts. This result contravened all our predictions."

The Soviets also thought it was a good idea to explode nuclear bombs near Earth. Happily, the result was a ban on such tests. Unhappily, the tests proved that if a rogue state did detonate even more powerful nuclear bombs in low Earth orbit it could make it unusable for satellites for years. A machine caught in the blast would be destroyed, and the ensuing radiation would fry any replacements.

All of these are realistic possibilities for future space wars. So what can be done to avert them?

Hawks among astropolitical thinkers are confident that because space militarization is happening, the way ahead is to escalate first, and to a level that competitors cannot match. This is a strategy of deterrence.

The age-old problem with arms control is that no one negotiates weapons limitations with someone who doesn't have weapons. The Thomas theorem was only coined by William Thomas and Dorothy Thomas in 1928, but it seems to have applied for all recorded history: "If men define situations as real, they are real in their consequences." Countries tend to define potential threats as real threats. Therefore, placing your bets on one of the space powers that decides not to match any military space advances by a rival is not recommended.

Military commanders are tasked by their political leaders to develop capabilities to promote what are considered the nation's interests. Take this example from Space Force's 2020 Planning Guidance document: "Space Force is called to organize, train, equip, and present forces capable of preserving America's freedom of action in space; enabling Joint Force lethality and effectiveness . . . Spacepower backstops deterrence by communicating America's ability to impose costs on hostile actors and deny adversary objectives." That's telling them. And telling them is part of the strategy.

There's a fine line between giving away your secrets and deterring your opponent by letting them know how strong you are. If you keep everything secret, the other side might think it can risk attacking. The arms reduction treaties between the Soviets and the Americans in the 1980s were underpinned by an agreement on joint inspections of their nuclear capabilities—"Trust, but verify," as Reagan said, although he borrowed the phrase from the Russian *Doveryai, no proveryai*.

Now American military space strategists debate whether the US should demonstrate to Beijing and Moscow the capabilities it has for destroying satellites in order to deter them from a surprise attack. Those in favor argue that you can't deter with invisible weapons. Those against say it could accelerate an arms race. The debate is as old as warfare. In the US Air Force it's called opening the "Green Door" because, as legend has it, there was an air force base where "above Top Secret" stuff happened, and the documents were behind a green door.

To date deterrence has prevented us from pushing the "big red button" because, according to Mutual Assured Destruction, every side knows a nuclear strike would result in retaliation, and we would all perish. As Everett Dolman explains, "MAD has three components; Mutual (everybody), Assured (no ifs or buts), and Destruction (total loss). If the threat is not credible . . . deterrence has failed."

But that doesn't stop us from engaging in more conventional forms of warfare. The same is true in space. While no one is reaching for the big guns—yet—there are still options available that won't destroy our ability to continue operating in space: jamming, spoofing, capturing, and hacking satellites without creating any appreciable debris, for example. And so MAD deterrence doesn't stop anyone from continuing to develop this sort of tech or engaging in lower-scale skirmishes—which could quite easily escalate.

The alternative is a burgeoning arms race. To counter this we need a series of comprehensive arms-control treaties.

Among the myriad threats, the biggest is probably competition between China and America and what is called in geopolitics the "Thucydides Trap." The term was popularized by the Harvard scholar Graham Allison in his book *Destined for War*. In it he quotes from Thucydides's *History of the Peloponnesian War*: "It was the rise of Athens and the fear that this instilled in Sparta that made war inevitable." For Athens read China, and for Sparta the US. Allison identified sixteen cases where a rising power threatened to displace an existing one and found that in twelve war was the outcome. In the four where conflict was averted, deployment of imaginative statecraft was required—for example, an intervention by the pope that resulted in the 1494 Treaty of Tordesillas, preventing a devastating war between Spain and Portugal, and more recently the US-Russia relationship that gave rise to the Cold War rather than nuclear bombs. In all four cases compromises were made, often messy ones with knock-on effects, but Allison's point is that they averted an all-out catastrophic military confrontation, and these examples can help the superpowers of the Space Age do the same. Compromise is now required by the Big Three.

There are many factors militating against this. China and Russia view American advances in space as designed to maintain the US's dominant position on Earth. In some respects they may be right. Similarly, the US remains anxious that technological achievements by the other two will be used to further military prowess that threatens America—and it too has a point.

It is difficult to know where to draw the line in terms of threat and counterthreat. For example, the Russians and Chinese both have a lead on new-generation hypersonic glide missiles. Unlike intercontinental ballistic missiles, which launch and fly on a predictable trajectory, the glide missile can maneuver through the upper atmosphere, changing direction and height at speeds above Mach 5—which is 1,715 meters per second. The Kremlin claims speeds well above this for its Avangard and Zircon weapons. The US missile defense systems cannot match these speeds in their response times, especially as without a consistent trajectory they don't know the target. Given that the warhead could carry a nuclear device, the temptation to assume a nuclear attack would be great, thus increasing the likelihood of unleashing a nuclear response before being hit.

As we've seen, the US is developing a layered defense against hypersonic missiles. It hopes to have sensors in space that can track them. At the same time the missile guidance system on board the attackers' satellites will be targeted from land, sea, and/or space. Further down the line, satellites capable of firing downward onto the missiles are likely.

There will also be the defense of commercial interests to consider. For centuries we have seen how the flag follows the trade. A recent example on Earth is the 2022 security agreement between China and the Solomon Islands, by which if Chinese interests on the islands are endangered (as they were during the 2021 riots that targeted Chinese property and people), Chinese government "forces" can come to their aid. States will take similar views of their commercial enterprises in space—the flag will follow the trade.

So, solutions. Professor Dolman has been arguing for a different course of action, proposing a Mutual Assured Reliance strategy: "Since space is inherently global—from an astropolitical analysis it is a single point in the cosmos—any benefit or loss that comes from it will be shared among all states; granted, not equally. Rather than focus on the fear of losing space access, instead we should make all states party to the gains to be made from the exploitation of space to create a green future for all humankind of abundance."

I'm sure most of us are with him all the way. It's just the getting there. Getting past the weapons-testing, the killer satellites, the probable military space stations and bases.

The twentieth-century French philosopher Raymond Aron may have died forty years ago, before some of our modern technological wonders, but even then he recognized our oldest problem: "Short of a revolution in the heart of man and the nature of states, by what miracle could interplanetary space be preserved from military use?"

Vive la révolution.

TOMORROW'S WORLD

When I dipt into the future,
Far as human eye could see;
Saw the Vision of the world,
And all the wonder that would be.
—Alfred, Lord Tennyson, "Locksley Hall," 1842

What was distant is now near, what was slow is fast, and the impossible is now the norm. With this in mind, our thoughts on space and the future should not be limited—not even, except on a practical basis, by science.

Contrast two beliefs. First Leonardo da Vinci: "I have always felt it is my destiny to build a machine that would allow man to fly."

And now the eminent Canadian American astronomer Simon Newcomb, who in 1902 said: "Flight by machines heavier than air is unpractical and insignificant, if not utterly impossible." The following year Orville Wright took off in the *Kitty Hawk*, also known as the *Wright Flyer*, and flew into the future that da Vinci had imagined.

We are now writing what will be history in space. We already have magnificent pioneers and amazing achievements. Where they did get to was incredibly hard.

The obstacles encountered in the next two decades will be

An illustration depicting the Mars helicopter Ingenuity and the Perseverance rover, both of which landed on Mars on February 18, 2021. Ingenuity was the first aircraft to successfully perform controlled powered flight on the Red Planet.

enormous, but if they are not overcome then we cannot progress to the challenges that lie beyond. Humanity has not got so far only to stand still now.

It will not all be "noble future of humanity" material. There's money to be made in space, and people are out to get it. The commercial opportunities are many. If spaceflight for ordinary people becomes the norm, then space hotels will not be far behind. Want your ashes scattered in low Earth orbit? There'll be a Galactic Funeral Service for that. If a company didn't mind outraging most of humanity, it could mar our night skies with horizon-wide advertising. If that doesn't float your spaceboat, then of more use might be new technology, such as Techshot's BioFabrication Facility, which it hopes will be used to print human organs in low Earth orbit, thus bypassing the problem of gravity encountered on Earth, the pressure of which restricts the natural growth of cells and tissues.

The first step toward this future will be made as we head back to the Moon. Many of the immediate problems we'll face once there are the same as those we have encountered for so long on Earth: food, water, shelter. But to them we must add the making of breathable air, and finding the power sources to do that, which needs to be done 238,855 miles from home.

The pioneers are already scouting the terrain. The early Apollo missions landed near the Moon's equator for many reasons, among them that on the journey home, in the event of a systems failure after liftoff, an equator launch allows a free-return trajectory—the craft loops around the Moon using its gravity and slingshots back to Earth.

The equator neighborhood is likely to be a prime location for energy, given that the regions most often exposed directly to the Sun probably have more concentrated deposits of helium-3 than the poles—and given that helium-3 has huge potential as a power source on the Moon, on Earth, and for further exploration (see chapter 3).

However, in the late 2020s and 2030s, the equator is probably not where the action will be. When you're looking for somewhere to live—think "location, location, location." Estate agents extolling the "excellent natural light" of a room, even when it's a coal cellar, might try to sell you a property near the Moon's equator using the same phrase. For two weeks there would indeed be constant natural light, but the following two weeks would be constant natural night. This is because a single rotation of the Moon takes about one Earth month, so a lunar day and night each last around fourteen Earth days. To put it another way—if you are gazing up from your pod on the Moon's equator, it will take the Sun 29.5 days to move all the way across the sky, disappear, and then return to its original position. This means that for half the time, even if you've gone to the Moon for a vacation, you will not be recharging your batteries, and on the Moon you need batteries.

But it's also true that equatorial temperatures swing from around 261°F during the lunar day to around -290°F during the lunar night, or, to put it in more scientific terms, from "Scorching hot!" to "Freezing the balls off a brass monkey." As you may be aware, the latter is an English idiom rooted in the myth that Royal Navy cannonballs were stacked in pyramids in a brass tray known as a monkey. When the temperature dropped dramatically, the brass contracted and the pyramid collapsed. It's not true. You'd hardly stack cannonballs in a pyramid, as they would roll around the deck each time the ship was hit by a wave. However, the idea of temperatures contracting and expanding metal is what counts—you really don't want the metal in your spaceship, oxygen cylinders, and living quarters to be expanding and contracting.

This is one of the reasons why the first machines and humans to visit the Moon always landed during the lunar dawn—at the beginning of the two-week-long lunar day—a time when the extremes of temperature, and temperature swings, could be avoided. Equipment could be designed to withstand either extreme heat or cold, but not the demands of immense temperature variations.

Given the difficulties of the equator region, the next craft are much more likely to land at the poles, which are considered the best place for permanent settlement. They are generally colder than the equator, but the temperature swing is much less severe, especially in the semipermanently lit areas.

As we've seen, the scientists are "house-hunting" in the South Pole–Aitken basin, where the Sun barely rises above the horizon and so cannot reach the depths of the craters. Therefore, most of them have been in shadow for billions of years and may contain the ice required to process into oxygen, water, and hydrogen—and that could mean rocket propellant that would make getting from the Moon base to Mars more viable.

NASA scientists have found several regions, all within six degrees' latitude of the pole, which are candidates for what they hope will be the first base. Each region is nine-by-nine miles and they contain multiple potential landing sites. The Sun is very low in the sky but there should be enough energy for the first settlers to harvest using solar panels, to power their way to a new beginning.

Given that breathable oxygen is something of a priority for living anywhere, fortunately there is a possible source for this too—the Moon's topsoil, called regolith. The term is used to differentiate between dust, soil, broken rock, and other materials on the surface and the solid rock underneath it. The impact of the constant meteorite bombardments that have hit the Moon for hundreds of millions of years can be seen clearly with a telescope costing a few hundred dollars. The surface is pockmarked with huge craters. What cannot be seen is the effects of the millions of micrometeorites that have left the topsoil sand-like, although the particles are much sharper and more abrasive than most sand found on Earth. Of course, regolith covers the whole surface, which means you don't have to go to the ends of the Moon to get it.

Bake regolith at a very high heat in a container, add hydrogen gas, plus a pinch of scientific knowledge, and water vapor forms that can be separated into oxygen and hydrogen. And . . . breathe.

And then breathe out—because astronauts' breath can also be tapped to produce oxygen, as indeed can their sweat and urine, using technology already developed for the ISS. As astronaut Douglas H. Wheelock told the *New York Times*: "Yesterday's coffee is tomorrow's coffee."

So, light, water, oxygen, energy—we're living off the land. All we need now is shelter. At first this is likely to consist of either flatpack or inflatable structures brought from Earth. They will have to be covered with regolith to protect the inhabitants from the huge amount of radiation permanently striking the Moon. Data from a German-built experiment on one of China's Moon missions suggests that, due to the lack of atmosphere, levels of radiation are two hundred times higher than on Earth's surface. Fortunately, regolith has a high resistance to solar radiation and a low thermal conductivity, which means it can be used as rough stucco for a Moon base.

Once that is up and running, other options can be explored, including a "basement apartment." The Moon has around two hundred pits that we know of, with caves leading off them, and in many the permanent temperature is 145.4°F—what the scientists refer to as "sweater weather." It's thought that overhanging rocks limit how hot the pits become during the day, and then prevent heat from dissipating at night.

A report published in the journal *Geophysical Research Letters* concluded that "Lunar caves would provide a temperate, stable, and safe thermal environment for long-term exploration and habitation of the Moon." Some are lava tubes similar to those found on Earth, where a river of lava has cooled leaving a long, hollow tunnel often with caves leading off the passageway. NASA and ESA astronauts are already in training to explore underground. Teams have been sent to the lava tubes on the Spanish island of Lanzarote to experience the terrain, practice commanding lunar rovers to drive in the tunnels, and construct 3D maps of the environment to assess its "traversability." There's an irony that so long after humanity left caves and began building, the most cutting-edge technology available will be used so that we can return to them.

When the water, oxygen, and energy sources are established, and the habitats and food greenhouses built, then attention will turn as quickly as possible to mining the Moon's abundant rare-earth elements.

This is all part of the rough template for the next ten years. Armstrong's "giant leap" is now being followed by a series of baby moonsteps that will lead to generations of humans born not of this planet. That is a long way down the line, and the challenges we must overcome to get there are numerous—not least protecting pregnant women from the dangers of radiation and low gravity—but the journey has begun.

So on to Mars. Launching from the Moon wouldn't put a dent in the vast distance between Earth and Mars but, as discussed, it does reduce the amount of fuel required. The planet contains all the problems to be encountered on the Moon, plus many more, and it's nearly six hundred times farther away on average. Putting humans on Mars is a far greater challenge.

For this trip timing is everything. It helps to set off during the period when the two planets are closest to each other; due to their elliptical orbits this happens every twenty-six months. If you want to catch that ride, you've just missed the closest approach in 60,000 years, which was in 2003. It won't be as close again until 2287.

If you had a car that could drive through space at about 62 mph it would take 256 years, and many an "Are we there yet?," before you arrived on Mars. If you had a spacecraft that could move at the speed of light, it would be a matter of minutes. Failing that, modern space probes launched from Earth so far tend to reach Mars in somewhere between 128 and 333 days, so you are going to have to allow for about nine months cooped up in a pressurized tin can. And if you want a return journey, you should block out two years because you'll have to wait a few months on Mars to make sure Earth is in the right place for the trip home. If you just took off again and continued your orbit around the Sun, when you got back to where you started Earth wouldn't be there. Which would be problematic.

In 2022, Mr. Musk pushed back the date of the first human landing on Mars to 2029. This is one of the years when the distance between Earth and Mars is reduced to around 60 million miles. That's quite a shortcut, given the average distance is about 140 million miles. If you're thinking of booking, the following dates might be useful for your house-selling plans and diary: May 2031, June 2033, September 2035, November 2037, and January 2040. If you want to be the millionth person to make the journey, try August 2050. Mr. Musk will celebrate his seventy-ninth birthday that year—possibly on Mars. Possibly not.

Mars is a "big ask." Any time someone gives a time frame for a crewed landing, add five years. Minimum. The internet is awash with articles from 2013/14/15 suggesting humans would arrive on the surface in the 2020s. The Dutch company Mars One took tens of millions of dollars from investors after saying it could land humans on Mars in 2023. It was declared bankrupt in 2019. NASA says 2033 is a "maybe" just for humans orbiting Mars and estimates 2039 for getting people to the surface. China has a reasonable timeline of between 2040 and 2060, but then it's always been good at taking the long view.

The most recent rovers have started exploring and mapping the surface of Mars. NASA's Curiosity rover has traveled about 19 miles since arriving in 2012. Perseverance has some catching up to do, but is pushing toward 9 miles following its deployment in 2021. They have been joined by China's Zhurong rover, and the ESA hopes to send its own in 2028. The UK-built Rosalind Franklin, named after the British DNA pioneer, had been due to launch on a Russian rocket in 2022, but the invasion of Ukraine put paid to that.

The first humans on Mars will probably have had the builders in before they move in. Robotic spacecraft will have done some of the heavy lifting, landing, and construction to enable the astronauts to carry more of what they need to survive. Another spacecraft could be positioned in orbit or on the surface with enough fuel to make it home, meaning the astronauts do not need to take huge amounts with them.

One of the problems the first settlers will face is that Mars is a bit chilly; at night temperatures drop to -76°F. Another is that we can't breathe there due to an annoying lack of oxygen. Of course we have methods of making it, as we plan to on the Moon, but that would restrict us to small-scale shelters, and wouldn't allow for a proper settlement of the planet. So, terraform it. "Nuke Mars!," as Musk tweeted in 2019. Explode nuclear bombs in order to release carbon dioxide and other gases stored in the soil and polar caps, create a greenhouse effect, and warm the planet—climate change as a good thing. Not all scientists agree that the surface contains anywhere near enough carbon dioxide to warm the atmosphere, and indeed some believe it would create a nuclear winter. But it's an idea, and, as Musk says, "Failure is an option."

Musk is an optimist. He's set himself the deadline of 2050 to have built a city on Mars for a million people. That's not a misprint. One million people.

The plan: He builds a thousand of his reusable Starships. Once the first pioneers have set up the basic infrastructure, you buy a ticket, take a ride, and get a job on the Red Planet. Musk is on record as saying his goal is to reach a ticket price of around the average value of a house. Homeowners might well sell up to afford it. After all, the chances of returning are somewhat smaller than if you relocate from Albuquerque to Denver. Musk has acknowledged this. He suggested that ads to sell tickets might bear resemblance to those Ernest Shackleton is said to have issued for his exploration of Antarctica: "Men wanted for Hazardous Journey. Small wages, bitter cold, long months of complete darkness, constant danger, safe return doubtful. Honour and recognition in case of success."

Musk says there is a 70 percent chance that in his lifetime a rocket will take him to the self-sustaining city on Mars he envisions. It's hard to believe, but kudos to Musk: for all his faults, he dares to dream. As he says, "Life can't just be about solving problems. There have to be things that inspire you, that move your heart." He also came up with a great line: "I would like to die on Mars. Just not on impact."

What Musk and his fellow settlers will need is a way to keep fit en route. There are numerous health issues associated with lengthy missions undertaken in the absence of gravity. In the short term there is "space sickness." Symptoms include vomiting, dizziness, disorientation, and even hallucinations. This usually passes after a few days, but the long-term problems get worse with each passing week in zero gravity.

Fluids make up about 55 percent of female human weight and 60 percent of male and, due to gravity, tend to accumulate in the bottom half of our bodies. Humans have spent the last few hundred thousand years walking upright, and so we have evolved systems to ensure that enough blood flows to the heart and brain when we are standing. Evolution is not going to be deterred by a few months in space, and so the systems continue to work even in the absence of gravity. But the result is increased fluid in the top of our bodies, which is why astronauts have swollen faces. A bigger problem, though, is that without gravity the heart does not have to pump as hard, which leads it to weaken. The same is true of all your muscles, which start to waste away. A weaker heart means a decrease in blood pressure, which in turn can reduce the flow of oxygen to the brain—not ideal at any time, but especially bad if you're involved in rocket science.

With no weight on them our bones also weaken and become brittle, especially the load-bearing ones such as those in the lower spine and hips. After just six months in space it can take astronauts' bones up to three years to recover.

This is why we see astronauts on the ISS using exercise machines. A swimming pool would be useful, albeit a little cumbersome, and the water might not cooperate. A gym is smaller, but even so that's a lot of extra weight. The problems would also occur on Mars, but to a lesser extent. The planet's gravity is about 38 percent of that on Earth.

Closer to home, Musk's space rival Jeff Bezos has his own ideas. He is working on what he calls "long-range problems"—namely that Earth will run out of energy supplies. His solution, as we've seen, is to move to cities in space. Inspired by the Princeton University physicist Gerard O'Neill's book *The High Frontier*, Bezos envisions mile-wide, sealed, wheel-shaped rotating cities stationed near Earth. This would allow millions of people to live in them, while other structures would house heavy industry, thus relieving Earth both of people and pollution. He accepts that the technology required is, at best, decades away, but says his company will begin building the infrastructure now. His space exploration company, Blue Origin, says it is on track to launch a commercial space station in the second half of this decade, which will house up to ten people in an area of 30,000 cubic feet.

His space cities will need to rotate in order to create artificial gravity to combat the numerous health hazards of extended stays in low- or zero-gravity environments. For example, it's doubtful a woman could have a normal pregnancy in space, hence Mars One, prior to filing for bankruptcy, was advising its prospective early settlers against trying to get pregnant once they arrived. So rotating crafts are a must—that's why we see such constructions in films like *The Martian* and *2001: A Space Odyssey*.

But not so fast! Not so fast that it impacts the fluid in the inner ear and induces nausea and disorientation. That means spinning at a leisurely 1–2 rpm, which requires a spacecraft of at least 3,281 feet in length. By no coincidence, both China and NASA are conducting feasibility studies for just that. Both know they are probably several decades from achieving it—after all, it took ten years to build the ISS—but they have their eyes on the horizon.

They might be helped by recent developments—such as dispensing with rocket fuel and engines and going back to the age of sail. A little over four hundred years ago the genius that was Johannes Kepler wrote: "Given ships or sails adapted to the breezes of heaven, there will be those who will not shrink from even that vast expanse." In 2004, two large solar sails were deployed into space by the Japan Aerospace Exploration Agency (JAXA).

It was Space Age origami. JAXA packed intricately folded panels into a small rocket that took off from the Uchinoura Space Center on Kyushu, Japan's third-largest island. It then released two sails, one in a clover-leaf shape 33 feet across, the other like a pleated fan, each ten times thinner than a sheet of paper. The Japanese proved that large, ultra-light structures can be folded and released intact. Several countries are now working on prototypes of larger, thinner models made of reflective heat-resistant materials that will act as solar panels and propel spaceships vast distances at incredible speeds.

We know sunlight exerts enough force to move objects: as light particles (photons) hit the sails, they push them forward. Constant sunlight equals constant propulsion equals constant acceleration, eventually up to five times the speed of a traditional rocket. NASA scientists liken this to the "Tortoise and the Hare." Launch a rocket and a sailcraft at the same time and the rocket will . . . rocket ahead. But the sailcraft will gradually accelerate to over 62 million mph, whereas the fastest rocket-propelled craft so far is the Parker Solar

Probe, which approached 430,000 mph. Putting it another way, one managed 0.064 percent of the speed of light, the other should be able to get up to 10 percent.

For an idea of the distances that could be covered, at that sort of speed you would be able to fly from Earth to the Moon in a matter of seconds. It's a work in progress.

In theory, such technology could eventually be used to propel humans across our solar system. However, given the difficulties involved, some might ask: Why not just keep sending robots? The question has been posed by, among others, the eminent astrophysicists Donald Goldsmith and Martin Rees. In 2020, they wrote an article titled "Do We Really Need to Send Humans into Space?" and summed up their answer with the subtitle "Automated spacecraft cost far less; they're getting more capable every year; and if they fail, nobody dies."

Nicely put. They point out that since the first Moon landing, hundreds of probes have been sent out across the solar system, visiting all the Sun's planets, and that machines could have done most of the scientific experiments carried out on board the ISS. They recognize the emotional pull of the heroics men and women accomplish in space and are not against researching alternative places for humans to live, but they come down on the side of safety and practicality and believe robots can achieve this.

Their argument is strongest when it comes to government budgets being spent on human space travel as opposed to private enterprise funding it. I'd argue that both governments and companies should be spending money and sending humans for several reasons. It's probable we will need a refuge from Earth at some point, and it's definite that we already need more resources to raise living standards here. There will be scientific, medical, and technological advances as we make this journey even if we don't yet know what they are, and now is not the time to press the pause button.

Yes, robots can, and should, do a lot of this, but they cannot tell us how it feels out there and what it's like psychologically to be so far from Mother Earth. Without the human factor, without inheritors of the mantle of Marco Polo, Ibn Battuta, Zheng He, Columbus, Roald Amundsen, Gagarin, Armstrong, and others, it will be harder to convince people that this is our future, and that the work done now is akin to the old adage that you plant a tree so that future generations can sit in the shade. Everything in our history tells us we cannot resist the call of the unknown. It is inevitable we will venture farther because, as US astronaut Gene Cernan put it, "Curiosity is the essence of human existence."

On to the distant future, where things get weird. Technologies such as space sails might seem fantastical, but television and walking on the Moon would once have been in that category. There are other possibilities that are currently in the realm of science fiction, but still merit looking at in theory.

Perhaps the most scientifically sound is the idea of space elevators. They were first proposed in 1895 by our Russian friend Konstantin Tsiolkovsky, whom we met in chapter 2. He envisaged a tower extending from Earth's surface up to geosynchronous orbit, which rotates at the same speed as Earth. You could then send things up in an elevator. Simple. In the twenty-first century the theory of space elevators is proven. It's just a case of finding the materials, the willingness—and the funding. The fact that even now we have not invented materials capable of supporting the weight of a tower 21,748 miles high does not detract from the visionary genius of a man who was thinking these things through before the first airplane had taken off.

Modern versions include starting from Earth and building upward; starting from the Moon and dangling a cable down toward Earth via a Lagrange point; or bypassing Earth and building a cable from a Lagrange point to the Moon. The advantage of the first two is that payloads could be lifted into space without the need for large rockets, and so the costs of space travel would be hugely reduced. Depending on which report you read, materials that might be used include 3.3-feet-thick steel cables or carbon polymers such as Zylon. Personally, I'd use either spider silk or that other strongest material known to humankind—chewing gum. Either way, if it's going to happen, and it's one of the more feasible scenarios, then securing the "tether sites" on Earth, the Moon, or the Lagrange points will be a prime target for future national security agencies.

Alternatively, when it comes to spacecraft, there's always good old warp factor 4.5, which, as numerous websites curated by earnest folk will tell you, is the average cruising speed of the starship *Enterprise* in *Star Trek*. There's a problem with the idea of warp factor anything. It's called Einstein's Special Theory of Relativity and the impossibility for anything to move faster than light. Warp factor 1 is the speed of light, so Einstein would have had conniptions at the thought of warp factor 7 reaching 343 times the speed of light. Which is quite fast.

Happily, theoretical physicists are not going to let the musings of the greatest scientific mind of the twentieth century get in the way. The theory is that the *Enterprise* does not travel faster than the speed of light; instead, it sits within a compressed "warped" bubble of space-time that is traveling faster than light. When this bubble arrives at the desired location, out you pop and surprise the Klingons. Hundred-meter (328 feet) sprinters would benefit from this sort of thing. Compress the 100-meter lane in front of you to 10, and you'll arrive at the finish line a lot more quickly than your rivals.

Off we go then. Except it seems it's slightly more complicated than that. Just one of the many problems is that it involves using enormous amounts of antimatter—that's the same stuff as ordinary matter except it has the opposite electric charge. An electron, which is one of the particles that makes up ordinary matter, has a negative charge. Its partner in these matters is a positron, which has a positive charge.

When antimatter collides with normal matter an explosion is produced, emitting pure radiation, which travels from the epicenter of the explosion at the speed of light. Unfortunately, there's not much antimatter around. Fortunately, we can make our own. The high-energy particle colliders (atom smashers), such as the one at CERN (the European Organization for Nuclear Research), create antimatter. Unfortunately, CERN only produces one to two picograms of the stuff each year. A picogram is a trillionth of a gram. This is enough to power a 100-watt light bulb for about three seconds, which, given that it would take tons of it for interstellar travel, is—in scientific terminology—"not very much." But getting to Mars might merely take a millionth of a gram, and NASA believes it may be only decades from this being achieved.

Of course, there's always wormholes, which would mean theoretically you could travel a vast distance almost instantly, arriving practically as soon as you set off. There is an analogy that provides a simplified idea of how this theory works: Two people hold a folded bedsheet, leaving a gap between the two layers. Place a bowling ball on the top half of the sheet and it will roll to the middle, causing the sheet to curve. Now imagine an equal force on the underside of the lower half of the sheet, causing that side to curve upward. In theory, if the force exerted on both sides was strong enough, it would create a passage joining these two separate places, potentially light-years away from each other, allowing a short, quick journey between them.

Weird enough? Finally, to teleportation. In 1998, some very, very clever people at the California Institute of Technology (Caltech) scanned the structure of a photon (a particle of energy carrying light) and then sent the information across a 3.3-foot coaxial cable where the photon was replicated. They also confirmed the theory that by so doing, the original photon was destroyed. This is because scanning the original disrupts it so much that it disappears, leaving only the copy existing wherever it has been sent. This essentially means that if we ever get to the stage where we can teleport humans, each time we do we will kill the original person but replicate them in another place. Again, and again.

Physicists specializing in quantum science have built on Caltech's breakthrough. Rearchers in China teleported a photon 60 miles in 2012, and followed up in 2017 by sending one to a satellite hundreds of miles above the Earth's surface; however, copying the octillions of atoms in a human body and sending the information to another planet seems a little way off. Studies suggest that even if we could teleport someone it would take the entire power supply of the UK for more than a million years to accomplish it; energy prices being what they are, who would even start? However, work is being done on sending quantum packets of information across thousands of miles. China has already beamed such information to its satellites in space. The prize here is a communication system that would be incredibly hard to hack; and, crucially, even if it was hacked, the transmitting party would know because "observing" anything in the quantum world causes it to change. A very rough analogy is that when you check your car's tire pressure you alter the pressure even if only by a fraction. If you cared enough to have a highly sensitive computer monitoring your tires, you'd know if someone had checked them.

This shows how the seemingly impossible might start to become a reality. We could go on. What about the probability of millions of variations of life existing on other planets? Numerous exoplanets—those beyond our solar system—have been identified as potential candidates for supporting life. As astrophysicist Neil deGrasse Tyson says about our current ability to see what's out there: "Claiming there is no other life in the universe is like scooping up some water, looking at the cup, and claiming there are no whales in the ocean."

We could spend our days speculating about the unknowable, the wonders, the fun stuff. But amid all the dreaming and theories, first we must rise to meet the challenges we already face: the arms race, the competition for territory and resources, the lack of laws, and many other negative aspects of this new era and domain we find ourselves in.

The space team at the giant investment management company Morgan Stanley points out the transformative effect that technological advances can have. It uses the example of the first demonstration of a safety elevator in 1854. Few people could foresee the impact it would have on the design of cities, but within two decades every multistory building in New York was built around a central elevator shaft and architecture rose ever higher. It believes the development of reusable rockets may be a similar turning point in the space industry. The lower cost of entry to space, including the reusable rockets pioneered by SpaceX, will accelerate investment and Morgan Stanley estimates that the industry will generate revenues of over $1 trillion by 2040, up from around $450 billion in 2022.

This could help humanity toward the goal of net-zero emissions on Earth. Technologically deploying "fields" of solar panels in space is already achievable. They could collect enough energy from the Sun to meet all current electricity needs and beam it down. Placing factories in space will be possible, and, as discussed, mining the Moon and asteroids for rare-earth materials and other resources is within our grasp.

Given all recorded human history, it is unlikely that we will recognize our common humanity and work together in space to harvest its riches and then distribute them equally, but even with nation-states and blocs competing with one another there will be common benefits for all. The probability that we will project into space our current concept of sovereignty, in which nation-states have power over mutually recognized territories, should not hold us back from our destiny as a species.

Stephen Hawking gets (almost) the last word. "Spreading out may be the only thing that saves us from ourselves. I am convinced that humans need to leave Earth." Enjoy the ride.

EPILOGUE

The past is the beginning of a beginning,
and all that is and has been is
but the twilight of the dawn.
—H. G. Wells

We've always had a sense of restlessness; it seems rooted in our genetic makeup. We wanted to see what was at the top of the mountain. We had an urge to sail out into the ocean. Once we'd fully mapped our terrestrial confines, it was inevitable that the moment we could go farther, we would.

We used to measure distance by how long it would take to walk from one place to another, then to ride on an animal, to drive, to fly. Now we're moving on to a different level of mathematics, with light speed and more zeroes than the average calculator can cope with. Some people argue that technology has negated geography, but in space all it has done is change the equation. Perhaps, though, the magnitude of the universe will prove vast enough for humanity to move past its history of power struggles and rivalry. As astronomer Carl Sagan said, "If a human disagrees with you, let him live. In a hundred billion galaxies, you will not find another." Perhaps.

What is certain is that we will continue to venture ever farther from Earth. We will settle on the Moon. We will live on Mars and beyond. It will take time, but we will find technological accelerators that will drive changes we cannot yet imagine. As science fiction writer Arthur C. Clarke put it: "They are as much beyond our vision today as fire or electricity would be beyond the imagination of a fish." But that must not deter us from forging ahead. Generation after generation of civilizations began to build great monuments knowing that they would not live to see them completed. Their legacy says: "This is what we did when we were here. It was for us, and it was for you."

Sputnik, Apollo, Soyuz, the ISS, and now Artemis and Orion are among the Space Age's great monuments. Future generations will look back on them and know that without them, and without Pythagoras, Newton, Tsiolkovsky, Gagarin, and Armstrong, they would not be wherever it is they are.

Perhaps by then they will have been able to peer behind the very first second of our 13.8-billion-year journey and find . . . something rather than nothing. All the imagined and unimagined wonders are out there, in front of us, waiting to be discovered by Homo spaciens.

ACKNOWLEDGMENTS

Thank you to Professor Everett Dolman, Dr. Bleddyn Bowen, Sangeetha Abdu Jyothi, Aardman Animations, Air Vice-Marshal Paul Godfrey, Professor John Bew, the UK National Space Centre, and those in the diplomatic and intelligence worlds who generously gave their time and knowledge but prefer to remain anonymous.

As always, thanks to the whole team at Elliott & Thompson: Lorne Forsyth for the freedom to write what I want, Jennie Condell and Pippa Crane for making it readable, and Amy Greaves and Marianne Thorndahl. And many thanks for the continuing support and hard work of the team at Simon & Schuster US.

SELECTED BIBLIOGRAPHY

African Union Commission. "African Space Strategy: For Social, Political and Economic Integration," October 7, 2019. https://au.int/sites/default/files/documents/37434-doc-au_space_strategy_isbn-electronic.pdf.

Ancient Origins. www.ancient-origins.net.

Bowen, Bleddyn E. *Original Sin: Power, Technology and War in Outer Space*. London: Hurst, 2022.

Bowen, Bleddyn E. "Space Is Not a High Ground," SpaceWatch. Global, April 2020. https://spacewatch.global/2020/04/spacewatch-column-april/.

Brunner, Karl-Heinz. "Space and Security—NATO's Role." Science and Technology Committee, NATO Parliamentary Assembly, October 10, 2021. https://www.nato-pa.int/download-file?filename=/sites/default/files/2021-12/025%20STC%2021%20E%20rev.%202%20fin%20-%20SPACE%20AND%20SECURITY%20-%20BRUNNER.pdf.

Brzeski, Patrick. "'Wandering Earth' Director Frank Gwo on Making China's First Sci-Fi Blockbuster," *Hollywood Reporter*, February 20, 2019. https://www.hollywoodreporter.com/movies/movie-news/wandering-earth-director-making-chinas-first-sci-fi-blockbuster-1187681/.

Brzezinski, Matthew. *Red Moon Rising: Sputnik and the Rivalries That Ignited the Space Age*. London: Bloomsbury, 2007.

China National Space Administration. http://www.cnsa.gov.cn /english/.

China National Space Administration. "Joint Statement Between CNSA and ROSCOSMOS Regarding Cooperation for the Construction of the International Lunar Research Station," April 29, 2021. http://www.cnsa.gov.cn/english/n6465652 /n6465653/c6811967/content.html.

"China's Film Authority Hails 'The Wandering Earth,'" *Global Times*, February 22, 2019. http://en.people.cn/business /n3/2019/0222/c90778-9548796.html.

Chow, Brian G. "Stalkers in Space: Defeating the Threat." *Strategic Studies Quarterly* 11, no. 2 (2017). https://www.airuniversity .af.edu/Portals/10/SSQ/documents/Volume-11_Issue-2/Chow.pdf.

Communist Party of the Soviet Union. Central Committee Presidium Decree. "On the Creation of an Artificial Satellite of the Earth," August 8, 1955. Wilson Center Digital Archive. https://digital archive.wilsoncenter.org/document/cpsu-central-committee -presidium-decree-creation-artificial-satellite-earth.

David, Leonard. "Is War in Space Inevitable?," Space.com, May 11, 2021. https://www.space.com/is-space-war-inevitable-anti-satellite -technoloy.

Doboš, Bohumil. "Geopolitics of the Moon: A European Perspective." *Astropolitics* 13, no. 1 (March 2015): 78–87. www.doi.org/10.1080 /14777622.2015.1012005.

Dolman, Everett. "Geostrategy in the Space Age: An Astropolitical Analysis." *Journal of Strategic Studies* 22, nos. 2–3 (1999): 83–106.

Eisenhower Library. "Reaction to the Soviet Satellite: A Preliminary

Evaluation," Memo to White House Staff, October 15, 1957. https://www.eisenhowerlibrary.gov/sites/default/files/research/online-documents/sputnik/reaction.pdf.

European Space Agency. "International Space Station Legal Framework." https://www.esa.int/Science_Exploration/Human_and_Robotic_Exploration/International_Space_Station/International_Space_Station_legal_framework.

Foust, Jeff. "Defanging the Wolf Amendment," *The Space Review*, June 3, 2019. https://www.thespacereview.com/article/3725/1.

Gillett, Stephen L. "L5 News: The Value of the Moon," National Space Society, August 1983. https://space.nss.org/l5-news-the-value-of-the-moon/.

Goh, Deyana. "The Life of Qian Xuesen, Father of China's Space Programme," SpaceTech Asia, August 23, 2017. https://www.spacetechasia.com/qian-xuesen-father-of-the-chinese-space-programme/.

Goldsmith, Donald, and Martin Rees. "Do We Really Need to Send Humans into Space?," *Scientific American*, March 6, 2020. https://blogs.scientificamerican.com/observations/do-we-really-need-to-send-humans-into-space/.

Goldsmith, Donald, and Martin Rees. *The End of Astronauts: Why Robots Are the Future of Exploration*. Cambridge, MA: Belknap Press, 2022.

Government Office for Science (UK). "Satellite-Derived Time and Position: A Study of Critical Dependencies," January 30, 2018. https://assets.publishing.service.gov.uk/government/uploads/system/uploads/attachment_data/file/676675/satellite-derived-time-and-position-blackett-review.pdf.

Grid Assurance. https://gridassurance.com.

Gwertzman, Bernard. "US Officials Deny Pressure on Paris to Go into Chad," *New York Times*, August 18, 1983. https://www.nytimes.com/1983/08/18/world/us-officials-deny-pressure-on-paris-to-go-into-chad.html.

Hayden, Brian, and Suzanne Villeneuve. "Astronomy in the Upper Palaeolithic?," *Cambridge Archaeological Journal* 21, no. 3 (September 2011): 331–55. www.doi.org/10.1017/S0959774311000400.

Haynes, Korey. "When the Lights First Turned On in the Universe," Astronomy.com, October 23, 2018. www.astronomy.com/news/2018/10/when-the-lights-first-turned-on-in-the-universe.

Hendrickx, Bart. "Kalina: A Russian Ground-Based Laser to Dazzle Imaging Satellites," *The Space Review*, July 5, 2022. https://www.thespacereview.com/article/4416/1.

Hilborne, Dr. Mark. "China's Space Programme: A Rising Star, a Rising Challenge," *China in the World*, Policy Series 2020, Lau China Institute. https://www.kcl.ac.uk/lci/assets/ksspplcipolicyno.2-final.pdf.

Horvath, Tyler, Paul O. Hayne, and David A. Paige. "Thermal and Illumination Environments of Lunar Pits and Caves: Models and Observations from the Diviner Lunar Radiometer Experiment," *Geophysical Research Letters* 49, no. 14 (July 2022). https://doi.org/10.1029/2022GL099710.

House of Commons Defence Committee. "Defence Space: Through Adversity to the Stars?," Third Report of Session 2022–23, October 19, 2022. https://committees.parliament.uk/publications/30320/documents/175331/default/.

"Jodrell Bank Lovell Telescope Records Luna 15 crash," YouTube. www.youtube.com/watch?v=MJthrJ5xpxk.

Kaku, Michio. *The Future of Humanity: Terraforming Mars, Interstellar*

Travel, Immortality, and Our Destiny Beyond Earth. London: Penguin Random House, 2019.

Kameswara Rao, N. "Aspects of Prehistoric Astronomy in India," *Bulletin of the Astronomical Society of India* 33 (2005): 499–511. https://www.astron-soc.in/bulletin/05December/3305499-511 .pdf.

Khan, Zulfqar, and Ahmad Khan. "Chinese Capabilities as a Global Space Power," *Astropolitics* 13, no. 2 (2015): 185–204. www.doi .org/10.1080/14777622.2015.1084168.

Korenevskiy, N., "The Role of Space Weapons in a Future War," Central Intelligence Agency, September 7, 1962. https://www .cia.gov/library/readingroom/document/cia-rdp33 -02415a000500190011-3.

Li, Chunlai, Chi Wang, Yong Wei, and Yangting Lin. "China's Present and Future Lunar Exploration Program," *Science* 365, no. 6450 (July 2019): 238–9. www.doi.org/10.1126/science.aax9908.

Maltsev, V. V., and D. V. Kurbatov. "International Legal Regulation of Military Space Activity," *Military Thought: A Russian Journal of Military Theory and Strategy* 15, no. 1 (January–March 2006).

Massimino, Mike. *Spaceman: An Astronaut's Unlikely Journey to Unlock the Secrets of the Universe.* New York: Crown Archetype, 2016.

Morgan Stanley. "Space: Investing in the Final Frontier," July 24, 2020. https://www.morganstanley.com/ideas/investing-in-space.

Morning Consult. "National Tracking Poll #210264, February 12–15, 2021." https://assets.morningconsult.com/wp-uploads /2021/02/24152659/210264_crosstabs_MC_TECH_SPACE _Adults_v1_AUTO.pdf.

Mosteshar, Sa'id. "Space Law and Weapons in Space," *Oxford*

Research Encyclopedia of Planetary Science (May 23, 2019). https://doi.org/10.1093/acrefore/9780190647926.013.74.

NASA. "Sputnik: The Beep Heard Round the World, the Birth of the Space Age," podcast, October 2, 2007, https://www.nasa.gov/multimedia/podcasting/jpl-sputnik-20071002.html.

NASA. The Artemis Accords, October 13, 2020. https://www.nasa.gov/specials/artemis-accords/img/Artemis-Accords-signed-13Oct2020.pdf.

NASA. "Memorandum of Understanding Between the National Aeronautics and Space Administration and the United States Space Force," 2020. https://www.nasa.gov/sites/default/files/atoms/files/nasa_ussf_mou_21_sep_20.pdf.

National Archives. "President John F. Kennedy's Inaugural Address (1961)." Milestone Documents. www.archives.gov/milestone-documents/president-john-f-kennedys-inaugural-address.

North Atlantic Treaty Organization. The North Atlantic Treaty, Washington D.C., April 4, 1949. https://www.nato.int/cps/en/natolive/official_texts_17120.htm.

NPP Advent. "Presentation on a Mobile Laser System to Shoot Down Drones." https://ppt-online.org/928735.

Oberg, James E. "Yes, There Was a Moon Race," *Air & Space Forces Magazine*, April 1, 1990. https://www.airandspaceforces.com/article/0490moon/.

Office of the Historian. "Letter from President Kennedy to Chairman Khrushchev, June 21, 1961," *Foreign Relations of the United States, 1961–1963*, Volume VI, *Kennedy-Khrushchev Exchanges*. https://history.state.gov/historicaldocuments/frus1961-63v06/d17.

Oughton, Edward J., Andrew Skelton, Richard B. Horne, Alan W. P. Thomson, and Charles T. Gaunt. "Quantifying the Daily

Economic Impact of Extreme Space Weather Due to Failure in Electricity Transmission Infrastructure," *Space Weather* 15, no. 1 (January 2017): 65–83. www.doi.org/10.1002/2016SW001491.

Parliament of Australia. "Ministerial Statement to the Parliament of Australia by Minister for Defence Mr. Stephen Smith," June 26, 2013, Hansard P7071. https://parlinfo.aph.gov.au/parlInfo/search /display/display.w3p;query=Id%3A%22chamber%2Fhansardr %2F4d60a662-a538-4e48-b2d8-9a97b8276c77%2F0016%22.

Parliamentary Office of Science and Technology. "Military Uses of Space," Number 273, December 2006. https://researchbriefings .files.parliament.uk/documents/POST-PN-273/POST-PN-273.pdf.

Public Opinion Foundation (FOM) Russia. "On the State and Development of the Space Industry and the Desire to Fly into Space." https://fom.ru/Budushchee/14192.

Reesman, Rebecca, and James R. Wilson. "The Physics of Space War: How Orbital Dynamics Constrain Space-to-Space Engagements," Center for Space Policy and Strategy, October 16, 2020. https://csps.aerospace.org/sites/default/files/2021-08/Reesman _PhysicsWarSpace_20201001.pdf.

Richard Nixon Foundation. "7.24.1969–Apollo 11 Astronauts Return from the Moon," July 24, 2011. https://www.nixonfoundation .org/2011/07/7-24-1969-apollo-11-astronauts-return-from-the -moon/.

Royal Australian Airforce. "Defence Space Strategy." https://www .airforce.gov.au/our-work/strategy/defence-space-strategy.

Sagan, Carl. *Billions and Billions: Thoughts on Life and Death at the Brink of the Millennium.* New York: Random House, 1997.

Sagan, Carl. *Cosmos.* New York: Random House, 1980.

Sankaran, Jaganath. "Russia's Anti-Satellite Weapons: An Asymmetric Response to US Aerospace Superiority," Arms Control Association, March 2022. https://www.armscontrol.org/act/2022 -03/features/russias-anti-satellite-weapons-asymmetric-response -us-aerospace-superiority.

"SBSS (Space-Based Surveillance System)," eoPortal. https://www .eoportal.org/satellite-missions/sbss#sbss-space-based -surveillance-system.

Silverstein, Benjamin, and Ankit Panda. "Space Is a Great Commons. It's Time to Treat It as Such," Carnegie Endowment for International Peace, March 9, 2021. https://carnegieendowment .org/2021/03/09/space-is-great-commons.-it-s-time-to-treat-it -as-such-pub-84018.

South African Astronomical Observatory. www.saao.ac.za.

SpaceX. "Mars & Beyond: The Road to Making Humanity Multi-planetary." www.spacex.com/human-spaceflight/mars/.

Space Café Podcast Archive. Spacewatch.Global. https://spacewatch .global/space-cafe-podcast-archive/

The State Council Information Office of the People's Republic of China. "China's Space Program: A 2021 Perspective," January 2022. https://english.www.gov.cn/archive/whitepaper/202201/28 /content_WS61f35b3dc6d09c94e48a467a.html.

Statista Research Department. "Government Expenditure on Space Programs in 2020 and 2022, by Major Country," Statista. https:// www.statista.com/statistics/745717/global-governmental-spending -on-space-programs-leading-countries/.

"Tactical Lasers," GlobalSecurity.org. https://www.globalsecurity.org /military/world/russia/lasers.htm.

United Nations Digital Library. Treaty on Prevention of the Place-ment of Weapons in Outer Space and of the Threat or Use of

Force Against Outer Space Objects. Draft texts submitted by the Russian Federation and the People's Republic of China, February 12, 2008. https://digitallibrary.un.org/record/633470?ln=en.

United Nations Office for Outer Space Affairs. Treaty on Principles Governing the Activities of States in the Exploration and Use of Outer Space, including the Moon and Other Celestial Bodies, December 19, 1966. https://www.unoosa.org/oosa/en/ourwork /spacelaw/treaties/outerspacetreaty.html.

U.S. Agency for International Development (USAID). "USAID Safeguards Internet Access in Ukraine Through Public-Private-Partnership with SpaceX," Office of Press Relations, April 5, 2022. https://www.usaid.gov/news-information/press-releases /apr-05-2022-usaid-safeguards-internet-access-ukraine-through -public-private-partnership-spacex.

U.S. Air Force Ballistic Missile Division. "Military Lunar Base Program: Volume 1," April 1960. https://nsarchive2.gwu.edu /NSAEBB/NSAEBB479/docs/EBB-Moon03.pdf.

U.S. Space Force. "Chief of Space Operations' Planning Guidance," November 9, 2020. https://media.defense.gov/2020/Nov/09 /2002531998/-1/-1/0/CSO%20PLANNING%20GUIDANCE .PDF.

Vidal, Florian. "Russia's Space Policy: The Path of Decline?," French Institute of International Relations, January 2021. https://www .ifri.org/sites/default/files/atoms/files/vidal_russia_space_policy _2021_.pdf.

Weeden, Brian. "2007 Chinese Anti-Satellite Test Fact Sheet," Secure World Foundation, updated November 23, 2010. https:// swfound.org/media/9550/chinese_asat_fact_sheet_updated_2012 .pdf.

The White House. "United States Space Priorities Framework," December 2021. https://www.whitehouse.gov/wp-content /uploads/2021/12/United-States-Space-Priorities-Framework -_-December-1-2021.pdf.

The White House. "Fact Sheet: Joined by Allies and Partners, the United States Imposes Devastating Costs on Russia," February 24, 2022. https://www.whitehouse.gov/briefing-room/statements -releases/2022/02/24/fact-sheet-joined-by-allies-and-partners -the-united-states-imposes-devastating-costs-on-russia/.

Whitehouse, David. *Space 2069: After Apollo: Back to the Moon, to Mars, and Beyond.* London: Icon Books, 2021.

Wilford, John Noble. "Russians Finally Admit They Lost Race to Moon," *New York Times*, December 18, 1989. https://www .nytimes.com/1989/12/18/us/russians-finally-admit-they-lost -race-to-moon.html.

Zhao, Yun. "Space Commercialization and the Development of Space Law," *Oxford Research Encyclopedia of Planetary Science* (July 30, 2018). https://doi.org/10.1093/acrefore/9780190647926.013.42.

INDEX

INDEX

INDEX